PHYSICS

D0191264

Also by Tony Rothman:

A Physicist on Madison Avenue

Censored Tales (UK)

Science à la Mode

Frontiers of Modern Physics

The World Is Round

INSTANT

PHYSICS

FROM ARISTOTLE TO EINSTEIN, AND BEYOND

BY TONY ROTHMAN, PH.D.

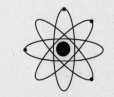

A Byron Preiss Book

FAWCETT COLUMBINE • NEW YORK

To my students, known and unknown.
The known, who demonstrate with crushing
directness that it is impossible to be too
clear. The unknown, who suffer in silence.

ACKNOWLEDGMENTS

In literature, the conflict between style and substance is eternal, but in science, there is no question that ideas are more important than words. Though the words in *Instant Physics* are, I trust, my own, I have benefited from the ideas of many colleagues. In particular I would like to thank Profs Sheldon Glashow, Roy Glauber, and Robert Kirshner for the opportunity to help teach their core courses at Harvard, where more than a few of the ideas discussed here surface regularly. Most of all, a thanks to my many students, past and present, to whom I have dedicated the book.

Cartoon Credits: © 1995 David Cipress—200. © 1995 Frank Cotham—222. © 1995 Leo Cullum—154. © 1995 Robert Mankoff—235. © 1995 John O'Brien—130. © 1995 Jack Ziegler—139.

The Cartoon Bank, Inc., located in Yonkers, NY, is a computerized archive featuring the work of over fifty of the country's top cartoonists.

Illustration Credits: © 1995 Archive Holdings—8, 16, 21, 26, 43, 54, 60, 73, 96, 122.

Interior Illustrations by Dmitry Kushnirsky.

Library of Congress Catalog Card Number: 94-94571

ISBN: 449-90697-3

Cover design by Pixel Press

Manufactured in the United States of America

First Edition: February 1995

10 9 8 7 6 5 4 3 2 1

CONTENTS

LESSONS FROM PHYSICS

YOU MUST REMEMBER THIS

Don't memorize, understand.

ENDS

If you are a potential victim of this book, leafing through its pages while obstructing traffic in the science aisle of your favorite bookstore, there is a good chance you once took a physics course in the vanished days of your youth. Perhaps you now want a refresher or perhaps, feeling guilty at your ignorance of the central role played by science in modern society, you merely wish to become a more cultured individual. In any case, your knowledge of physics is inexcusably weak and, if you have indeed been exposed to the subject before, you probably regard the word "physics" as synonymous with "plague," as in "to be avoided like the."

This is a misconception. The word "physics" actually comes from the Greek *physika*, meaning nature, or more precisely, natural things. As we use it, physics refers to that sublime branch of science dealing with nature at its most fundamental level. Thus, the behavior of atoms, gravity, electromagnetism all fall within the domain of physics; the latest sightings of Elvis do not.

Physics is best thought of as the search for the rules that govern the behavior of the universe. Inspired by this great tradition, the plan of *Instant Physics* is to trace the moments of epiphany (look it up) that have led physicists to discover the most important of the universal rules. As you joyfully tread the path toward Enlightenment, you will become a True Believer in rules. The most fundamental are the famous laws of nature, which may be regarded as the physicist's version of the Ten Commandments: "Thou shalt not exceed the speed of light." "Thou shalt conserve energy." "Thou shalt increase in entropy," etc. But the laws of nature are both stricter and kinder than the Ten Commandments. If you break one of the commandments of Moses, you suffer mere eternal damnation. The laws of nature are much stricter—you cannot break them. On the other hand, their certified invincibility saves you from suffering the consequences.

Let us pause for a moment of philosophical reflection. Breathe deeply. In order for physicists to discover the laws of nature, the rules of the universe, they must assume such rules exist. The very assumption that there *are* rules is perhaps the most important thing that distinguishes science from other modes of thought, for example, politics. By their very nature, laws place limits on what is allowed and what isn't. You may have encountered a physicist at a cocktail party scoffing at UFOs because "faster-than-light travel is impossible." Or at perpetual motion machines because "they violate the laws of thermodynamics. Impossible." Indeed, "impossible" ranks number two in the physicist's lexicon (after "nonsense"). Such an attitude may explain why physicists do not often appear at cocktail parties.

The view that some things are impossible is indeed not a popular one, yet it is perhaps the most important message of the natural laws. It shall be emphasized throughout the short lifespan of *Instant Physics*. You aren't convinced? It is too pessimistic. No, it is reality. Still hesitant? Still believe in warp drives and perpetual motion machines? Then you have no choice but to read the following pages. With the Laws as our shepherd, you will see the error of your ways and be gently guided toward Truth.

But is it truly possible that *Instant Physics* harbors the keys to esoteric knowledge and worldly success? Not the latter. To be honest, "instant physics" is an oxymoron. As much as you might like, you cannot nuke physics in a microwave oven for seven minutes, swallow it whole and digest it. Unlike many subjects, English literature for example, physics is cumulative. You must start from the beginning. During its long training, a physicist repeats the same subjects over and over again, each time at a more exalted level, until after a decade of training the new researcher is unemployed. Thus our first

Guiding Principle: To know physics is to do physics.

To elaborate, the goal of this book cannot be to give you a working knowledge of physics. If you want a working knowledge of physics, go back to school; the results have just been indicated. The more modest goal of *Instant Physics* is to give you an idea of why physicists believe what they believe and to train you to use this sacred knowledge in the war against the infidel.

MEANS

By what paths shall we attain Enlightenment? For your dining convenience *Instant Physics* has been sliced lengthwise. The body of the text is indeed physics-lite. Those who hyperventilate when confronted by an equation will be pleased to know that no more than a handful of baby equations appear in the text proper. Instead, emphasis has been placed on principles and discoveries, served with a delightful historical seasoning and personality parade. The topics have been chosen for their scientific importance, as well as because they conveniently appear in several survey courses the author has helped teach at Harvard; they are stamped with the Ivy League Seal of Approval. In the spirit of reductionism and CNN Headline News, a "Sound Bite Summary" (here, "Summary") has been included at the end of each chapter, with one or two Key Words, a Key Definition, and the Key Concept. All attempts have been made to limit the Key 'cept to one sentence in order not to overtax memories and crib sheets.

A word about the historical approach. A typical physics text says, "So-and-so discovered this, so-and-so proved that, in such-and-such a year." This is physicist history, which corresponds to what a modern physicist using modern, well-established concepts assumes its illustrious forebears accomplished. Physicist history never corresponds to real history. Real history is much more complicated, and much more instructive. Science does not move forward by relentless logic alone, and sometimes it moves backward. There are inspired guesses, which are outnumbered by the great mass of lousy guesses; at the time of a discovery scientists often don't know what they're doing; for every person who receives recognition, ten do not, and so on. These observations may be summed up in the

> **Principle of Literary Oversight:** Textbooks may be straightforward and succinct, but the path of science is crooked and tortuous.

In a tiny book like this, much must be ignored altogether (China, the Arabs). Thus we cannot delve into real history. We will endeavor to indicate it exists.

The physics-lite reader will notice that the first few chapters, in particular, contain a fair dose of philosophy. This is inescapable. Physics as we know it grew out of philosophical ideas, and even now is strongly influenced by them. Indeed, it was not until about a century ago that physics became known as physics; before that it was a branch of natural philosophy. As you will see, philosophical ideas often hindered the progress of physics, and only by careful observation and experiment were these ideas discarded. An important theme, therefore, will be a corollary to the **fundamental principle of science** in Chapter 1:

The Physicist's Code: A single good observation is worth a century of bad philosophy.

Now, we have said the book is sliced lengthwise. For connoisseurs of real beer, accompanying the lite text is a series of demonstrations that contain *MATH*. The lite reader may ignore the demos completely. However, although students often approach teachers after class with the tearful plea that they are interested in "concepts," not math, to the physicist the distinction is not obvious. While admitting that physical concepts are not necessarily mathematical ones, the natural language of physics is mathematics and many of the laws cannot be expressed precisely in English, or even Sanskrit.

So, for those armed with high-school algebra, the demos highlight some of the most important concepts encountered in a basic physics course. We use nothing more than addition, subtraction, multiplication, and division—promise! (well, maybe the occasional taking of a square root)—and some grey matter to derive many exciting results. Those who are willing to sit down with a pencil and paper during commercial breaks and copy out the derivations will be pleasantly surprised at the amazing things that can be achieved, like $E = mc^2$ even. Understanding that difficulty lies in the eye of the beholder, your guide along the Shining Path has marked each demonstration with one to three exclamation points: !—mild; !!—spicy; !!!—extra hot.

In order to digest the more exotic dishes, the gourmet reader must nail down the concepts in order of appearance. This requires, in turn, understanding *exactly* what each vocabulary word means. This point cannot be overstressed. When a physicist utters the word "force," it means one and only one thing: a push or a pull, quantitatively a mass times an acceleration. When laycreatures use "force," it may mean power or equally well refer to

the Dark Side. "Energy," while it may accurately describe what you lack, these days has something to do
with harmonic convergence. Despite what Humpty Dum-
pty said, words have meanings.

An important moral lurks here. Many laycreatures believe that to know impressive vocabulary is to know physics. False! Utterly, unbearably false! To know definitions—
correct definitions—is not enough. Why? Behold: Knowledge of physics resides not so much in the concepts as
in the *connections* between the concepts. One can learn
all the definitions and theorems and proofs, but unless
one knows how they relate to each other—*and how they
do not relate to each other*—one is lost. As you cut through
the thicket toward Enlightenment, it will help to bear in
mind the

> **Principle of Multilateralism:** Concepts in physics
> are often related. Physics is internally consistent; no
> concept may violate any other concept.

But also remember

> **Separability of Issues:** Not all concepts are the same;
> not all concepts are applicable to a given problem.

To take a cue from Gertrude Stein: a force is not energy
is not momentum is not power. . . . In order to help you
keep issues separated, in each chapter you will find an
Esoteric Terms feature. Each is a treasure-trove of terms
defined as they shall be used.

Finally, never forget the **Prime Directive:**

> *Don't memorize, understand.*

And **mantras:**

Process is more important than results.

Creativity is more important than knowledge.

If, after finishing this book, the principles become part of your world outlook, *Instant Physics* will have served its purpose. Now, onward, Enlightenment awaits! Let's rock and roll.

SHADOWS OF FORGOTTEN ANCESTORS

IT'S ALL GREEK TO ME

YOU MUST REMEMBER THIS

You only arrive at the right answer after making all possible mistakes. The mistakes began with the Greeks.

Physicists Revere Their Elders

Socrates, Plato, Aristotle: To laycreatures, the great trinity of Greek philosophy. Socrates is the wisest of men, Milton tells us. Cicero would rather be wrong with Plato. Faustus would live and die in Aristotle's works.

Yet try a simple experiment: Unearth a physicist; cleverly find an excuse to praise these illustrious Greeks; observe the results. It is possible your guinea pig will merely mutter, "Philosophy. Humph. Never read much," and walk away. Otherwise, the odds are high that you will see the wan smile of a mild-mannered academic transform into a scowl, then into a murderous frown; the face will redden, the eyes turn heavenward. And then the torrent comes, a veritable diatribe, none of which you can follow or care to, but which in essence translates into one thing: Why are you wasting your time reading Plato and Aristotle, these criminals who set back the course of civilization a thousand years? Once the physicist calms down, it might tell you to burn your Plato and investigate some names that mean next to nothing to you: Thales, Anaximander, Empedocles, Parmenides. . . .

"The pre-Socratics," the physicist will say unequivocally, and the discussion ends. Why, when virtually all college philosophy courses focus on the Big Three, does the physicist spurn them and instead honor a handful of individuals known to us only by their names and thoughts? To understand the reason for this cultural schism let us ask one of the primordial questions: What is the nature of the world? This is the basic question of science, and people began asking it long before the ancient Greeks.

But what is interesting here is not so much the question as the answers. In all ancient civilizations, the answer was couched in terms of gods and goddesses. For in-

WHO'S WHO

Thales of Miletus (c.625–c.545 B.C.)
Made a splash by suggesting that water is the essential substance of the universe.

Anaximander of Miletus (c.610–c.545 B.C.)
Defined the primary ingredient of the cosmos as *apieron,* or "the indefinite." Vague, but eerily contemporary.

Pythagoras of Samos (fl. c.530 B.C.)
Cult leader who saw numbers as the underlying foundation of reality. Tortured generations of sixth graders, via the Pythagorean theorem.

Parmenides of Elea (fl. c.480 B.C.)
Believed that all being is eternal and that change is therefore impossible. A bad pick to run a self-improvement seminar.

Empedocles of Acragas (fl. c.440 B.C.)
Said the universe stems from four root elements: fire, air, water, and earth. Returning to his roots, he leaped into a volcano.

Aristarchus of Samos (fl. first half of third century B.C.)
Originated the heliocentric hypothesis, according to Archimedes, but didn't pursue it. So much for the Aristarchean Revolution.

Archimedes of Syracuse (c.287–212 B.C.)
A man of many togas—mathematician, inventor, physicist. Discovered the Archimedes principle, an early version of calculus, and the phrase "Eureka!"

stance, according to the ancient Egyptian creation myth, in the beginning were the Primordial Waters, Nun, boundless, changeless. In one version of the story,[1] from Nun arose the primeval Hill, Atum, "the Complete One." Atum, being alone and containing everything, was forced to create the world by unorthodox means. Sometimes he is said to have spit out the deities Shu (air) and his sister Tefenet (moisture). Other times Atum, having no consort, is said to have mated with his own hand. In either case, to Shu and Tefenet were born Geb and Nut, the earth and sky. From this point the other gods follow, in a more orthodox—if incestuous—manner.

Similar creation myths were of course told everywhere around the world. By the early Greeks, too. Hesiod in his *Theogony*, written about 700 B.C., tells a creation story that might have been plagiarized from the Egyptians. Homer, probably somewhat earlier, populates the *Iliad* and *Odyssey* with gods and goddesses whose meddling is largely responsible for the whole train of fiascos. (Homer, by the way, introduced the first scientific unit of measurement: the milliHelen, or beauty required to launch one ship.) Herodotus, like many people today, regarded an eclipse as a supernatural omen. In all these early writings, natural phenomena are personified or associated with deities.

THALES DISCOVERS WATER

Suddenly, in the sixth century B.C. something changed. At the Ionian Greek city of Miletus, in what is now Turkey, there appeared a group of men who devoted themselves to useless questions. And dumb answers. It is these

---

[1]See Carmen Blacker and Michael Loewe, editors, *Ancient Cosmologies* (London: George Allen & Unwin, 1975).

physikoi, as Aristotle later termed them, who with their pointless speculation set the course Western science has followed to the present day.

The first of the *physikoi* of whom we have any record was Thales of Miletus. Thales is credited with measuring the height of a pyramid by the use of shadows. He supposedly engaged in astronomical observations and predicted the eclipse of 585 B.C. The ancient Chaldeans discovered that solar eclipses occur in the same location every 54 years, 34 days. This **saros cycle,** as it is called, makes it easy to predict eclipses. Almost certainly Thales would have relied on the saros cycle. Plato relates that "Theodorus, a witty and attractive Thracian servant-girl, is said to have mocked Thales for falling into a well while he was observing the stars and gazing upward; declaring that he was eager to know things in the sky, but what was behind him and just by his feet escaped his notice."[2] Thus Thales comes down to us as the first absentminded professor. Well, Plato never got anything right.

Be that as it may, Thales is mostly remembered for his question: What is the basic substance of the world? and his answer: water. Why Thales believed water to be the stuff from which all other stuff arises, we haven't the faintest idea, and of course his answer is nonsense. But of far greater interest is the idea that there should be one and only one primary element.

The search for the primordial stuff was diligently carried on by Thales's successors. Anaximander of Miletus, Thales's pupil or colleague—we don't know which—is mostly remembered for the idea that the primary substance is something he termed *apieron,* a word that nobody can pronounce but which means "the boundless" or "the indefinite." Apparently having observed that

---

[2]G. S. Kirk and J. E. Raven, *The Presocratic Philosophers* (Cambridge: Cambridge University Press, 1971), p. 78.

everyday substances change, he reasonably concluded that the underlying stuff must be something else. What exactly he meant by *apieron* is the stuff of debate, but he may have had in mind something both infinite in extent as well as undifferentiable. From *apieron* the whole universe arose. Whatever it may have been, the *apieron* has an eerily modern ring to it, as we shall see in Chapter 5.

And just to be sure disagreement was complete, other pre-Socratics thought fire was the original element. Or air. Hitherto undiscovered fragments also show that several pre-Socratics became convinced of the primacy of Coca-Cola, having observed that "where there's life, there's Coke."

What is the modern reader to take from this? A six pack? First, as already mentioned, the Ionians seemed intent on reducing all observed phenomena to the fewest possible phenomena—even one phenomenon. Just as important, we notice the *physikoi* forgot something in their explanations: any mention of supernatural causes or deities. It can thus justifiably be said that Thales and his followers introduced two important Guiding Principles into the method of inquiry, which physicists have pigheadedly followed ever since:

Reductionism: The world is made up of a few underlying principles.

Mechanism: The world operates like a machine. Supernatural causes have no place in science.

Plus Ça Change . . .

Time flies like an arrow. Fruit flies like a banana. Both are manifestations of change. The concept of change proved particularly indigestible—and fruitful—to the pre-Socratics. Parmenides of Elea, tackling a conundrum that remains with us today—or at least until Chapter 9—argued that it is impossible to conceive of something coming out of nothing ("for never shall this be proved, that things that are not, are." Read it again.). In doing so, he managed to convince himself that change is logically impossible, and with that sleight of hand, relegated the changing world around us to illusion.

Democritus and Leucippus, in an attempt to reconcile the logic of Parmenides with the obvious of daily life, argued that the world is composed of *atoms*, which are incorruptible, unchangeable, immutable, but which nevertheless move, collide, and congregate, and by this helter-skelter activity produce the change we see all around us. And so, Parmenides's speculations produced an idea that one thousand years later changed civilization.

Parmenides's thoughts on the impossibility of change also influenced Empedocles, who first announced that only fools (who have no far-reaching thoughts) can believe something came out of nothing, but who then reversed himself to become a thoroughly modern physicist by claiming that it is only through the senses that we come to understand nature. Still troubled by the notion of change, he argued that there are four elements—earth, air, fire, and water—which are unchanging, and all apparent change is due to motion of these, obviously induced by Love and Discord.

The idea of the four elements, plagiarized by Aristotle, confused science for nearly two thousand years. But that wasn't good enough for Empedocles. He went on to demonstrate that air is a real substance by noting that water cannot enter a submerged vessel until the air is let out. In doing so he proposed the first lecture-demo on record. Empedocles also conjectured that light traveled at a finite velocity, which today we know is absolutely true. All of which is not bad for a man who supposedly ended his life by jumping into a live volcano.

ENTER MATH. OH NO, PLEASE, NO, ANYTHING BUT THAT. . .

Pythagoras— We don't know what Pythagoras looked like, and he didn't invent the Pythagorean theorem, but here he is and where would geometry be without it?

While Parmenides and Empedocles were debating change, the Pythagoreans contemplated number. (Hey, everybody needs a job.) Of Pythagoras of Samos we know nothing except that he flourished around 530 B.C.; about his early followers we know even less. But by the early fifth century a mystery cult (something like the Hari Krishnas) had grown up around the legendary figure, a cult that believed the underlying reality of all things was "number." Whether they meant that material objects are actually composed of numbers or that everything can be described by number is unclear. In either case they had discovered numerous numerical relations that gave them faith in the power of numbers. "Number, number, Hari number . . ." For instance, the Pythagoreans knew the mathematics behind the basic musical intervals and swore oaths by the **tetractys of the decad.** The famous Pythagorean theorem on right triangles was actually known to the early Mesopotamians. Pythagoras's plagiarism, or independent discovery of it, thus provides our first example of a folk-saying in science that might be termed the

Marketing Principle: Either you do the calculation or you get the credit.

To put it another way, somebody else always did it first.

Tetractys of the Decad—
This figure shows at a glance that
4 + 3 + 2 + 1 = 10, and so the Py-
thagoreans swore their most binding
oath by it: "Nay, by him that gave
to our generation the tetractys, the
fount and root of eternal nature."

In any case, by assuming that the world can be described by mathematics, the Pythagoreans introduced two more Guiding Principles dear to all natural philosophers since:

The Tetractys Doctrine: The fact that mathematics describes the real world so well cannot be coincidental. Mathematics is the correct language of physics.

Quantification: You don't know anything unless you've measured it and assigned a number to it.

Whether these doctrines are entirely true is questionable. At your next cocktail party you may wish to confront a physicist with a few choice lines given in the box. Nevertheless, we see that the achievements of the pre-Socratics were considerable. In their world view *chaos*—the mythological disorder from which all else sprang—was replaced by *logos* (law) and *kosmos* (order and universe). It is the search for *logos* and *kosmos* that became the guiding doctrine of physics.

THE PLOT THICKENS

Unfortunately, Plato and Aristotle got in the way. All thrillers need villains; Plato and Aristotle fit the bill for

COCKTAIL PARTY CONVERSATION

A physicist, wearing a "Tetractys of the Decad" T-shirt, is surrounded by admirers. You, wearing an "Alchemists Anonymous" T-shirt, approach.

YOU: Do you really believe mathematics describes the real world?

PHYSICIST: Better than poetry.

YOU: Then why do science profs always say things like "assume the cow is spherical"?

PHYSICIST: Well, one must . . . um . . . simplify!

YOU: A spherical cow? Isn't math describing something that doesn't exist?

PHYSICIST: Hmm, by the tetractys, this cannot be denied.

YOU: On the other hand, I've heard that most equations describing real systems can't be solved exactly.

PHYSICIST: Well yes, we usually approximate.

YOU: So, how can mathematics be such a great description of the real world? On the one hand you can describe things that don't exist; on the other hand you can't describe things that do exist. Sounds a lot like poetry to me.

PHYSICIST: But we can make our approximations as close as we want.

YOU: Didn't Einstein say, "As far as the laws of mathematics refer to reality, they are not certain; and as far as they are certain, they do not refer to reality"?

PHYSICIST: Did he say that? Excuse me—it's time for some nectar and ambrosia.

(Physicist disappears into the crowd.)

physicists. Most physicists haven't actually bothered reading Plato or Aristotle and so their dislike of them may
be somewhat irrational. But it cannot be denied that
Plato deviated from the True Path. Mocking the mecha-
nistic explanations of the pre-Socratics as pure nonsense,
he reintroduced a divinity, the Demiurge, into explanation. He also advocated the famous "world of forms,"
an ideal realm where perfect prototypes of all imperfect
things on the Earth existed. The world of forms—Plato's
"real world"—lay beyond the realm of the senses and
was not accessible to experiment. It could only be com-
prehended by logic. This was a completely antiscientific
approach.

Not content to dismiss the advances of his predeces-
sors, Plato in his *Timaeus* then laid a curse on physics:
The universe was created as a perfect sphere and the
planets travel in perfect circles at constant speed. As we
will see in Chapter 1, the origin of modern physics was
largely a matter of getting over this idea.

Aristotle, Plato's student, said, "The principles of every
science are derived from experience." Would you buy a
used car from this man? Despite his pronouncements,
Aristotle clearly paid little attention to observation, pre-
ferring logic instead. More important, his notion of
causes appears very unmechanical: Things don't happen
because of chance or collisions of atoms, but because of
a purpose. "Nature abhors a vacuum" and "The natural
tendency of the Earth is to move to the center of the
universe" are two such statements. Perhaps the most notorious was: "Circles are the natural state of motion."

This doctrine inscribed the Platonic curse in blood
and in doing so set back the course of physics for a
millennium. To be fair, most of the damage was actually
perpetrated by the Neoplatonists, those followers of Plato
and Aristotle who not only signed in blood, but chiseled

the masters' words into stone and refrained from any critical thinking of their own. In any case, many students even today hold Aristotelian ideas. As we go along, check whether your mind belongs in the ancient or modern world. Don't worry—despite Parmenides, change is possible.

THE SILVER AGE

Luckily, Greek science did not end with Plato and Aristotle. The tradition of the pre-Socratics continued into the Hellenistic period, after about 300 B.C., and centered around the city of Alexandria, where most of the famous men trained or worked. The names Euclid and Archimedes are familiar to almost every student. Euclid, in his series *The Elements*, codified what was then known about geometry. Eudoxus of Cnidos and Archimedes of Syracuse may fairly be said to have invented a version of calculus over one thousand years before Newton and Leibniz. Apart from his wide-ranging mathematical work, Archimedes also discovered the principle that bears his name, and that can be used to measure the density of different objects. Eureka!

Archimedes, in a work called *The Sandreckoner*, mentions that most astronomers believe the Earth is the center of the universe. He then reports on the hypothesis of one Aristarchus of Samos, that "the fixed stars and the Sun remain unmoved, that the Earth revolves around the Sun in the circumference of a circle, the Sun lying in the middle of the orbit."[3]

Because Aristarchus's own writings on this topic have not survived, it is not known whether he had any observa-

_____

[3]Sir Thomas Heath, *Aristarchus of Samos* (New York: Dover, 1981), p. 302.

tions to back up his hypothesis. But there is no doubt that Aristarchus in the third century B.C. had invented the heliocentric, or Sun-centered, solar system, over one thousand years before Copernicus. Here is a prime example of the Marketing Principle.

Archimedes's colleague Eratosthenes made a great contribution to astronomical knowledge by using trigonometric methods to measure the circumference of the Earth. Physicist tradition has it that he came very close to the correct answer. Physicists forget that the size of his unit of measurement, the *stadium*, is uncertain, so it is impossible to say whether he was indeed close or was out of the ballpark. The important point, though, is that his method was correct and that the Greeks knew the Earth was round. In general, the great contribution of these Hellenistic scientists was that they were unafraid to use the sledgehammer of mathematics to crack the secrets of nature.

Unfortunately for science, having come close to discovering modern astronomy—and from there modern physics—the Greeks turned their backs on it. Aristotle was to blame. His insistence that the Earth was the center of the cosmos and only circular orbits were allowed for the planets led to the infamous Ptolemaic system of the universe, completed by Ptolemy of Alexandria in the second century A.D. It was this Earth-centered monstrosity that lasted until Copernicus rediscovered Aristarchus's system in the sixteenth century. Leaving the Greeks in decline, we turn to the Ptolemaic system now.

ESOTERIC TERMS
(es-ə-'ter-ik tərms)

- *Mechanism*—The idea that the universe behaves like a giant machine. You might say Rube Goldberg was a Mechanist.

- *Reductionism*—The conviction that the diversity of the world is a manifestation of a few underlying principles.

- *Geocentrism*—The ultimate in narcissism: the Earth is at the center of the universe.

- *Heliocentrism*—Penultimate narcissism: the sun is at the center of the universe.

- *Presliocentrism*—Elvis is at the center of the universe.

SUMMARY

Key Words: Reductionism, mechanism.

Key Definitions: Reductionism, mechanism.

Key 'cepts: Reductionism, mechanism.

CHAPTER 1

TOWARD A CLOCKWORK UNIVERSE

PUTTING THE EARTH IN ITS PLACE

YOU MUST REMEMBER THIS

A theory is accepted only when the last of its opponents dies off. The Copernican Revolution was a great shift in mankind's thinking, but did not take place overnight.

COPERNICUS SETS THINGS SPINNING

The Copernican Revolution was like a bomb with a slow fuse. Once it had been set in place, it lay unnoticed for decades. No one realized how dangerous it was. And then suddenly—boom!—it exploded. And when it did, it changed everything.

Copernicus—
Not the first to bring the Sun to center stage, but his timing was better than his Greek predecessor, Aristarchus. The universe would never be the same.

The revolution that once and for all put the Sun at the center of the solar system and led to the creation of modern physics did not receive an auspicious start. The book that shook the world, *On the Revolutions of the Heavenly Spheres* by Nicolas Koppernigk, was published on May 24, 1543, while its author lay on his deathbed. The clergyman had written it thirty years earlier, but being a secretive Pythagorean he believed, unlike most researchers, "perish, not publish." Once on the market, it became an all-time worst-seller—the first one thousand copies never even sold out. Most historians have never read it because it is unreadable. Consequently, much of what textbooks say about the work is nonsense. Books of course need not be read to cause scandals, but Copernicus was denied even that consolation. For fifty years *Revolutions* was soundly ignored. Then all hell broke loose.

To understand how this retiring canon reluctantly changed the world, we must return to the Greeks. As mentioned in the Prologue, Aristotle decreed that the Earth's natural place was the center of the universe and that the natural orbits of the planets were circles. Taking political correctness to extremes, medieval scholars cowed themselves into believing Aristotle was synonymous with Truth.

But there is a serious problem with putting the Earth at the center of the universe. You have probably never paid attention to the skies enough to notice what was obvious to everyone two thousand years ago: The planets behave in a very peculiar manner. Mars, for instance, usually travels slowly east to west against the background of stars. But sometimes it slows down, stops, and moves backward for several months before it resumes moving forward again. To explain such *retrograde motion* became one of the main tasks of astronomy. If one assumes a geocentric universe—one with the Earth at the center—Mars must really at times travel backward.

The Aristotelians were undeterred. Ptolemy of Alexan-

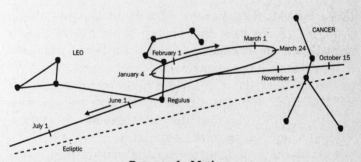

Retrograde Motion—
From the fall of 1994 to the spring of 1995, Mars will
move through the constellations of Cancer and Leo.
From January 4 through March 25, Mars will appear
to move backward. To explain this retrograde motion
was one of the central problems of astronomy.

dria managed to explain retrograde motion in a geocentric universe by placing the planets on a system of big wheels and little wheels that is best understood by referring to the below figure. Ptolemy termed the large wheels "deferents" and the small wheels "epicycles." Their combined motion caused the planet to sometimes move forward and sometimes backward, but to account for the motions of the five known planets he needed well over 40 of these invisible wheels.

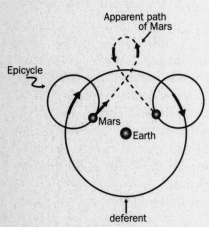

Apparent path of Mars

Epicycle

Mars

Earth

deferent

Deferents and Epicycles— Ptolemy tried to explain retrograde motion with a system of wheels within wheels. Mars rides on a little circle (the epicycle), which in turn rides on a big circle (the deferent). The whole thing became a hopeless mess. But the combined motion of the deferents and epicycles can make Mars appear to move backward.

It was a hopeless mess. In fact, to this day the phrase "adding epicycles" refers to a last-ditch attempt to patch up an already fatally flawed theory by the addition of further improbabilities. Tax reform is reminiscent of this procedure.

On the other hand, from the figure on page 19 you see that if one assumes a heliocentric system—one with the Sun at the center of the solar system—to explain retrograde motion becomes quite easy.

Now, in the Ptolemaic system the planets traveled in perfect circles but—in violation of Aristotle's decrees—not at uniform speeds. Copernicus, it turns out, was a committed Aristotelian. He could not tolerate such a heresy and set out to correct it. Eventually his fanaticism led to Aristarchus's heliocentric system. Yes, he knew of his prede-

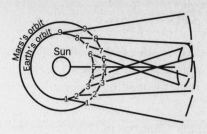

Heliocentric Explanation of Planetary Motion—
A simpler explanation for retrograde motion requires
putting the Sun at the center of the solar system.
Because the Earth is faster than Mars in its orbit
around the Sun, when the Earth overtakes Mars,
Mars appears from the Earth to move backward.

cessor and even mentioned him in his manuscript. But what did he do? Before publication he deleted Aristarchus's name, as well as those of his own assistant and teachers. Thus a corollary to the Marketing Principle:

> **Survival of the Loudest:** The top priority in science is priority. (Also known as the Zel'dovich Principle:[1] "Without publicity there is no prosperity.")

At any rate, contrary to what textbooks invariably say, Copernicus's commitment to Aristotle forced him to a model that actually contained more epicycles than Ptolemy's. What's more, the center of the solar system was not really the Sun but near it; strictly speaking, it was a quasiheliocentric theory. But since no one read the book, the details didn't matter. The important point was that now the Earth moved. Now the Earth was no longer the center of the universe. It may have been the most important conceptual shift in history. Who would have thought it all came from trying to explain why Mars moved backward?

---

[1]After a famous Russian astrophysicist, Yakov Zel'dovich.

**WHO'S
H
O**

Ptolemy of Alexandria (fl. A.D. 125–150)
Championed the geocentric model of cosmology, thus misleading everyone for 1,400 years. Ptoo bad.

Nicholas Copernicus/Nicolas Koppernigk (1473–1543)
Stole Aristarchus's idea that the Sun was the center of the solar system. Became famous. The universe never looked the same.

Tycho Brahe (1546–1601)
A more accurate astronomer than fencer. Provided best data in Western world to Kepler. Lost his nose in a fight; replaced it with a silver replica.

Johannes Kepler (1571–1630)
We should all be so crazy. Discovered three laws of planetary motion, which led to the creation of modern physics.

Galileo Galilei (1564–1642)
Built a telescope. Tangled with the church. Maybe dropped stuff off the Tower of Pisa. Discovered the correct relationships between distance, velocity, and acceleration, and the law of inertia (in fact, the inertia of the Italian bureaucracy still exists today). And you thought you were busy.

Isaac Newton (1642–1727)
One of science's greatest thinkers, he put it all together and invented mechanics. Is to physics what Moses is to the Old Testament: the bringer of laws, if not burning bushes.

KEPLER: METHOD IN HIS MADNESS

Johannes Kepler—He may have been a bit mad, but his three laws of planetary motion led to the creation of modern physics.

It was left to Copernicus's successors to foment the revolution. Johannes Kepler, born in Germany in 1571, is one of the most fascinating figures in the history of science. He was a practicing astrologer, held all sorts of mystical beliefs, made extraordinary mistakes in his investigations—and yet nevertheless managed to discover the three laws of planetary motion, which led to the formation of modern physics.

Kepler believed with Copernicus that the Sun lay at the center of the solar system, but wanted to know why the planetary orbits had the sizes they did. In a flash of inspiration he decided the orbits lay within the five "Platonic solids." The idea was totally nuts (and cheating didn't make it fit the data) but he believed that without publicity there is no prosperity and so published anyway in 1596. His book, called the *Mysterium Cosmographicum*, is not so important for its conclusions, which were nonsense, but for the transition from the medieval to the modern. It contained two concepts that have influenced physics ever since. Kepler wrote:

> What we have so far said served merely to support our thesis by arguments of probability. Now we shall proceed to the astronomical determination of the orbits and to geometrical considerations. If these do not confirm the thesis, then all our previous efforts shall have doubtless been in vain.[2]

---

[2]Arthur Koestler, *The Sleepwalkers* (New York: Macmillan, 1959), p. 255.

That is, after beginning from a crazy premise, he then says that if the results are not confirmed by observation, the theory is worthless. With this statement Kepler made one of the earliest and clearest enunciations of the

> **Fundamental Principle of Science:** All scientific statements must be testable by observation. As Confucius say: "Without the data, yo' chatta don' matta."

Kepler also assumed in the *Mysterium Cosmographicum* that there was a **force** emanating from the Sun "spreading in the same manner as light" that kept the planets moving. This was a revolutionary assumption. Before Kepler, no one thought to look for **causes** to explain the motion of the planets. Copernicus's system, as Ptolemy's, was merely a geometric construction. If your construction didn't work, just add another wheel. It's much like claiming the budget deficit increases interest rates; if that doesn't work, say the opposite. But Kepler tried to find a **force**, a mechanism that would keep the planets moving. The idea that a force was something that keeps objects in motion came from Aristotle (who decreed the natural state of motion was rest) and it was absolutely wrong. It would require Newton to straighten things out. Yet a new and important principle had been introduced into natural philosophy:

> **Causality:** Things happen for a reason. Those reasons are mechanical; they involve forces that can be measured. If the causes are duplicated, so are the effects.

A lesser scientist might have been content to let the *Mysterium Cosmographicum* stand, complete with its mad ideas and imperfect agreement with observations. But Kepler, crazy as he was, knew more work needed to be done.

He knew he needed observations and he knew where to get them: from the greatest astronomer of the age, Tycho Brahe.

TYCHO BRAHE: NO NOSE FOR DATA

Tyge Brahe, or Tycho as we know him, led what might be understated as a colorful life. Born to a noble Danish family in 1546, he was kidnapped by his childless uncle, a vice admiral, when his father refused to turn him over as once promised. Sometime later the vice admiral, having saved the king from drowning, died of pneumonia, and Tycho's fortunes abruptly increased. But God giveth and God taketh away. As a student Tycho fought a duel with another noble over who was the better mathematician and lost part of his nose. In portraits he is seen to sport a gold and silver replacement and so bears a slight resemblance to Oz's tin woodsman.

For science, what was important is that in 1576 the king bestowed upon Tycho an entire island with all its subjects. On this island Tycho built the fabulous Observatory Uraniburg and spent twenty years making the most precise astronomical observations the Western world had known. Yet, Tycho was not a creative genius and realized he could not wring the secrets of nature from his data. But he knew who could.

Kepler and Tycho met near Prague in 1600. These were the days of the counterreformation, which was soon to turn into the Inquisition, and Kepler, a Lutheran, was fleeing Germany for his life. Tycho had fallen into disfavor with the Danish crown. In this case we must be grateful for religious persecution and politics, for if the two had not met, physics as we know it might never have been born. Ah, a new principle:

Ivory Burns (think Zen).

KEPLER'S THREE LAWS
OF PLANETARY MOTION
. . . .

1. **The planets travel in ellipses with the Sun at one focus.**

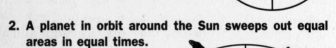

> *Kepler's Law #1*
> *All planets orbit the Sun in an*
> *ellipse, having the Sun at one focus.*

2. **A planet in orbit around the Sun sweeps out equal areas in equal times.**

> *Kepler's Law #2*
> *The equal area law says*
> *that a planet in orbit*
> *around the Sun sweeps out*
> *equal areas in equal times.*
> *The planet moves faster near the Sun, and so in a given*
> *time it travels a greater distance than when it is far from*
> *the Sun. But near or far, the areas swept out in that*
> *time are the same.*

3. **The square of the period of a planet's orbit is proportional to the cube of its average distance from the Sun.**

> *Kepler's Law #3*
> *On the vertical axis is shown*
> *the cube of the nine planets'*
> *average distances from the*
> *Sun. On the horizontal axis is*
> *shown the square of all the*
> *planets' periods. When one is*
> *plotted against the other a*
> *straight line results, showing*
> *that the two quantities are*
> *proportional to each other.*

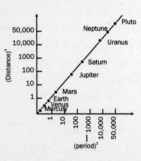

4. **Extra! Law Concerning Daily Planetary Motion: Clark Kent may never be seen in the same room as Superman.**

Tycho and Kepler had what might be best described as a love-hate relationship. It ceased abruptly after eighteen months, when Tycho died, but by then Kepler had what he wanted: the data. He worked on them through 1605, at which point he wrote his epoch-making *Astronomia Nova*, which was published four years later. In the *New Astronomy*, Kepler states:

> I am much concerned with the investigation of the physical causes. My aim in this is to show that the celestial machine is to be likened not to a divine organism but rather to a clockwork ... insofar as nearly all the manifold movements are carried out by means of a simple magnetic force, as in the case of a clockwork all motions [are caused] by a simple weight. Moreover I show how this physical conception is to be presented through calculation and geometry.[3]

He didn't quite get it, but he got the first two laws of planetary motion. Years later, in 1618, after successfully defending his mother in a witch trial, he discovered the third, which he published in his *Harmonies of the World*. The three laws of planetary motion are highlighted in the box.

GALILEO INVENTS FRESHMAN PHYSICS (AND GETS TESTED)

To Kepler, his laws were beautiful geometric relationships discovered by trial and error. Otherwise he did not know what to make of them. His irascible contemporary, Galileo Galilei, simply refused to believe them. Galileo

---

[3]Gerald Holton, *Thematic Origins of Scientific Thought* (Cambridge: Harvard University Press, 1975) p. 72.

remained convinced that the planetary orbits were circles and in that he deviated from the True Path. But in other respects this professor of mathematics at the University of Padua fought bravely on the side of Enlightenment. His was perhaps the best scientific publicist of the age. When in 1609 he heard that the telescope had been invented, Galileo promptly built his own and turned it to the sky.

Galileo—
Got in trouble with the church for being irascible
and uncompromising. Italians also consider
him the founder of modern physics.

The result was the *Sidereus Nuncius*, or "Starry Messenger." In it he announced several astronomical discoveries, including that of four moons around Jupiter. Until then the Earth was the only planet with a moon. Galileo's discovery that the Earth was nothing special in this regard was an argument in favor of Copernicanism. The very existence of things that could be seen only through a telescope caused indigestion to the Aristotelians, who held that nature contained nothing that could not be perceived by the unaided senses. Also in 1610 Galileo discovered the phases of Venus. This was the most important of his astronomical discoveries because the pattern of phases (similar to those of the Moon) proved that Venus could not orbit the Earth and that the Ptolemaic system must be wrong.

As the number of attacks on the Ptolemaic system increased, its defenders rely more and more on religious arguments. The year is 1616. You are there.

CHURCH FORBIDS TEACHING COPERNICUS
Inquisition Warns Padua Prof,
Galileo Says, "Nolo Contendere."

But in 1632 Galileo publishes his famous *Dialogue on the Two Chief World Systems,* a thinly disguised defense of the heliocentric hypothesis. He is found guilty of "suspicion of heresy," the second most serious religious crime, and spends the last ten years of his life under house arrest. And that's the way it is.

In arresting him, the Inquisition may have immolated itself because Galileo used his leisure to write the *Dialogue on Two New Sciences,* which was more important for the future of physics than his astronomical work. In it he reported the results of his experiments on the motion of objects, experiments abandoned a quarter century earlier that have, at least among Italians, earned Galileo the title of founder of modern physics.

The importance of the *Two New Sciences* can hardly be overestimated. Here Galileo refuted the Aristotelian doctrine that a heavier object falls faster than a lighter object and showed that all objects fall at the same rate (see Demo 2). And the famous, although probably fabricated, image of Galileo avidly dropping all manner of objects off the Tower of Pisa has since endeared him to generations of gullible freshmen. Galileo also rejected Aristotle's notion that forces were necessary to keep objects in motion. From experiment he postulated the

> **Law of Inertia**: Objects not acted on by a force travel in straight lines at constant speeds. Or if they are at rest they stay at rest.

(Though Galileo still thought the natural motion of planets was circular.)

Let us pause. Many people today have the Aristotelian idea that you need to apply a force to something to keep it moving. *Au contraire.* A turtle ice-skating on a frictionless rink would sail on forever and ever and . . . The law of inertia says you need a force only to *change* an object's state of motion (like friction from the ice slowing down the skater). But anyone who has tried to budge the IRS knows what inertia is. The law of inertia is so important that Newton took it over as his first law of motion. Thou shalt remember it.

Galileo, though apparently for the wrong reasons, also arrived at the correct relationships among distance, velocity, and acceleration that all beginning students are forced to learn (see Demo 1).

NEWTON PUTS IT ALL TOGETHER

To Isaac Newton was left the task of synthesizing what had come before. So much has been written about Newton that it would be superfluous to do so here; in any case, he was an unpleasant, secretive man and we do not even know precisely when he arrived at many of his monumental results. We do know that in 1665, while he

DEMO 1

Distance, Velocity, and Acceleration à la Galileo !!

You will need these formulas in nearly all other demonstrations throughout the book.

Some of the most important formulas in a basic physics course are those due to Galileo that connect distance, velocity, and acceleration. We derive them here.

To begin, suppose you go on a trip from Boston to New York, roughly speaking 400 kilometers (km). You may have exceeded the speed limit at times, going 120 km/hr, and you may have been stopped by the police, during which time your velocity was 0 km/hr. But if you made the trip in 4 hours, you would say your average velocity was 100 km/hr. You got this simply by dividing 400 km by the time, 4 hours. In general we can say

$$d = v_{avg}t, \qquad (1)$$

where d really represents the *change* of distance (400 km) and t represents the *change* in time (4 hours). If the velocity is always the same, you just get

$$d = vt \ (\textit{for } v = \textit{constant}). \qquad (2)$$

This also tells us $v = d/t$, or

velocity = (change of distance)/(change of time).

These three expressions say basically the same thing. Note also that since $v = d/t$, it is measured in *kilometers per hour* or *meters per second*.

DEMO 1

(continued)

Onward. If you accelerate from 0 to 100 km/hr at a constant acceleration, it is pretty easy to believe that your *average* velocity is 50 km/hr, which is just $(0 + 100)/2$. In general, for constant acceleration,

$$v_{avg} = \tfrac{1}{2}(v_o + v_f) \qquad (3)$$

where v_o is the initial velocity and v_f is the final velocity. Plugging this expression for v_{avg} into (1) gives

$$d = \tfrac{1}{2}(v_o + v_f) \times t = \tfrac{1}{2}(v_o t) + \tfrac{1}{2}(v_f t). \qquad (4)$$

Now, just as velocity is the change of distance with time, acceleration is defined as the change of velocity with time:

acceleration = (change of velocity)/(change of time).

So, if you start from 0 velocity with a constant acceleration a, then just as in equation (2)

$$v = at. \qquad (5)$$

Note that since $a = v/t$ and v is measured in *meters per second*, a must be measured in *meters-per-second per second*, usually written m/s^2. Now, in (5) v is the final velocity after a time t, so we can call it v_f. Equation (5) assumed you started from 0 velocity. But if you start from a velocity v_o, then you must add this on and the final velocity is:

$$v_f = at + v_o. \qquad (6)$$

DEMO 1

(continued)

Plugging (6) for v_f into (4) gives

$$d = \tfrac{1}{2}(v_o t) + \tfrac{1}{2}(at + v_o)t,$$

or, adding the first and the last terms,

$$d = v_o t + (\tfrac{1}{2})at^2. \qquad (7)$$

Equations (2), (5), and (7) are used in physics every day and will be used throughout this book. The last is particularly important. It tells us how far an object goes under constant acceleration, starting from an initial velocity v_o. It often happens in a problem that v_o is 0, so (7) becomes simply

$$d = (\tfrac{1}{2})at^2. \qquad (8)$$

The important thing here is that *the distance is proportional to the square of the time.* For instance, suppose you drop a frog from the top of a building. The acceleration of gravity is 9.8 m/s^2 (nearly 10 m/s^2), so after 1 second, the frog falls about 5 meters, after 2 seconds 20 meters, until . . .

You can also invert (8) to get the amount of time it takes to travel a given distance at a given acceleration:

$$t = \sqrt{\frac{2d}{a}}. \qquad (9)$$

Then plugging (9) into (5) gives the formula

$$v = \sqrt{2ad} \qquad (10)$$

which will be important in Chapters 3 and 4. You've now completed your first semester of physics.

was a student at Trinity College, Cambridge, the university was shut down for two years because of the Black Plague. During this enforced recess he began his research into optics, differential calculus, motion, and gravity. On top of which, London had its Great Fire, and Newton himself was allegedly clobbered by an apple. So, scientific breakthroughs aside, it was tough times all around.

One great breakthrough was to come up with a precise and consistent definition of force. (Notice we have cleverly avoided defining it so far.) **A force in physics is merely a push or a pull.** I punch you in the mouth, I've exerted a force on you. You summon the Dark Side, you haven't exerted a force on me. Specifically, you have been at rest. When I punch you, you begin to accelerate. Your state of motion has been changed. (Do you know the definition of acceleration? If not, see Esoteric Terms.) According to Newton's second law the force I've exerted on you is then equal to your mass (in kilograms) multiplied by your acceleration. In symbols: $F = ma$. This is probably the most important equation in physics, certainly in this book, and THOU SHALT REMEMBER IT.

Another breakthrough was to realize that the same force—gravity—that pulled apples to the ground held the Moon in orbit, something that had evidently not occurred to either Kepler or Galileo. All Newton's results were published in 1687 in his *Principia Mathematica*, which many regard as the single most important event in the history of science. The contents of the *Principia* may be summarized by Newton's three laws of motion plus the law of universal gravitation. In modern language they are stated below.

It is impossible to overstate the richness of these laws. As a first step, Newton used them to derive Kepler's three laws of planetary motion. That is, using $F = ma$ and the law of gravity, he showed that Kepler's laws must

NEWTON'S LAWS OF MOTION!
. . . .

These laws are at the bottom of almost every physics problem. Nail them.

1. An object travels at a constant velocity unless acted upon by an outside force. (The law of inertia.)
2. The force acting on an object is equal to the product of the object's mass and acceleration. (F = ma.)
3. For every action there is an equal and opposite reaction.

The Law of Universal Gravitation:

4. The gravitational force between two objects is proportional to the product of their masses and inversely proportional to the square of the distance between them. ($F = Gm_1m_2/r^2$.) Physicists might also agree that some similar rules apply to couples engaged in reproductive activity.

follow. Thus, in one fell swoop Newton showed that the planets obeyed the same laws as everything else.

In doing so, Newton invented the branch of physics known as **mechanics**, the science that seeks to understand the behavior of objects as they interact among each other by forces. It may safely be said that the next two centuries of physics were an exploitation of Newton's laws. Here we remain content to point out three important features.

First, Newton's laws cannot be proven. The second law is a definition and the first and third may be considered hypotheses subject to experimental test. (Indeed, the third is not true under all circumstances.) Thus, with all other natural laws, Newton's laws have in common the

> **Axiomatic Property:** Natural laws are the axioms of physics. They cannot be proven; they can only be postulated and tested by experiment.

TALMUD (COMMENTARY ON THE LAWS)

Newton's laws are designed to catch the uninitiated with conceptual pitfalls. Beware! For True Enlightenment, one needs to understand the vocabulary employed. See Esoteric Terms for definitions.

1. The First Law.

Here, "constant velocity" includes zero velocity (rest). Thus the alternate form of the law frequently encountered:

An object remains at rest or in uniform motion unless acted upon by an outside force.

"Uniform motion" means motion in a straight line at constant speed (and has nothing to do with the law as interpreted by the California Highway Patrol). Now, clearly, the first law does not mean much until force is defined. As Arthur Eddington once quipped, the First Law states "an object remains at rest or in uniform motion insofar as it doesn't." The concept of force is made more precise in the Second Law.

2. The Second Law.

One sees that to define a force one needs in turn the concept of mass. The concept of mass, along with length and time, is left undefined in physics. We say a certain platinum cylinder housed in the Bureau of Standards in Paris has a mass of one kilogram (kg).

The above form of the Second Law is not quite correct. To be accurate, the Second Law states that the force of acting on an object is equal to the rate of change of the object's *momentum (mv)*. If the mass of the body is constant, then only its velocity is changing. Since by definition the rate of change of the velocity is the acceleration, we get $F = ma$, as above. But if the mass of the object is changing (as in a rocket), this must be taken into account.

TALMUD (COMMENTARY ON THE LAWS) (*continued*)
. . . .
3. The Third Law.

  What this means is that if you exert a force on me, I exert a force of the same magnitude on you, but in the opposite direction.

4. The Law of Universal Gravitation.

  In this expression, m_1 and m_2 are the two masses, r is the distance between them. G is a number called the gravitational constant, which fixes exactly how big F_{grav} is. G must be measured in the laboratory. When the m's are in kilograms and r in meters, G is found to be 6.67 $\times$ 10^{-11}. The important thing is that there is an r^2 in the denominator. That means if the distance is increased by a factor of 2, the force goes down by a factor of 4, etc. Many natural phenomena in nature obey such "inverse-square laws." *My Three Sons*, on the other hand, follows the "complete squares law."

DEMO 2

Gravity, Centrifugal Force, and the Moon !

You need the definitions of force, velocity, and acceleration for this demo. You will need the results of this demo for Demo 2, Chapter 8.

Drop a brick and a marshmallow simultaneously from the same height. They hit the ground at the same time. (Using a sheet of paper isn't fair because air resistance screws things up.) Amazing, isn't it? Actually, it is; this stupid observation proved the

DEMO 2

(continued)

basis for Einstein's general theory of relativity (Chapter 9).

But, but WHY? Here's why: According to Newton's law of gravity, the gravitational force between two objects is

$$F_{grav} = Gm_1m_2/r^2 \qquad (1)$$

as elucidated in the Talmud.

Suppose m_1 is the mass of the Earth, r is the radius of the Earth, and m_2 is you. Then we call F_{grav} your *weight*. Weight is a force.

Now, Newton's second law $F = ma$ is ALWAYS true. It is important to remember that in $F = ma$, a and m refer to the same object. For instance your weight is $m_{you}a_{you}$, not $m_{Earth}a_{you}$ or any other combination. Since $F = ma$ is always true, let's see what happens when we set it equal to your weight, F_{grav}:

$$F_{grav} = m_{you}a_{you} = Gm_{Earth}m_{you}/r^2_{Earth}. \qquad (2)$$

Notice your mass m_{you} miraculously cancels out from both sides of the equation and your acceleration is just

$$a = Gm_{Earth}/r^2_{Earth}. \qquad (3)$$

We have dropped the subscript "you" on the acceleration, because Equation (3) shows that the acceleration depends only on m_{Earth} and r_{Earth}, not on the mass of the falling object. Thus all objects fall to the Earth at the same rate, just as you and Galileo observed.

DEMO 2

(continued)

Now, if you put the mass and radius of the Earth (5.98×10^{24} km; 6.4×10^6 m) into (3), *a* comes out to 9.8 m/s², or about 10 m/s². This is the acceleration of gravity on the surface of the Earth; since it is so important, it gets a special letter *g*:

$$g = Gm_{Earth}/r^2_{Earth} = 9.8 \text{ m/s}^2. \quad (4)$$

One of Newton's great achievements was to connect (1) with the planets. How did he do this? Probably it had less to do with apples than with the Moon. The Moon is traveling around the Earth in a circle. Therefore a force must be acting on it (otherwise by Newton's first law it would travel in a straight line). Since there is a force acting on it, the Moon must be accelerating ($F = ma$!). In this case, although the *speed* of the Moon is constant, its *direction* is changing, and so by definition it is *accelerating*. All this is true for any object traveling in a circle—be it the Moon or a rat whirled around your head on a string. The inward force the string exerts on the rat, or gravity on the Moon, is known as **centripetal force**.

But for every action there is an equal and opposite reaction. The outward force the rat exerts on the string is called **centrifugal force**. (The same force you feel on one of those spinning amusement park rides. Remember: centrifuge.) Since centrifugal and centripetal forces are equal and opposite they have the same size (which is why nobody can keep them straight). The size of the centripetal force (centrifugal—whatever) is

DEMO 2

(continued)

$$F_{cent} = mv^2/r. \quad (5)$$

This is an important formula. (We can actually derive it—more or less—in Demo 3.) Here, r is the length of the string or the size of the orbit. But $F = ma$ (always!). That means

$$F_{cent} = mv^2/r = ma_{cent}.$$

Again the mass miraculously cancels out and so we have

$$a_{cent} = v^2/r. \quad (6)$$

This is the formula for the centripetal acceleration (centrifugal—whatever).

Just for the hell of it, let's figure out what a_{cent} is for the Moon. The distance to the Moon is about 380,000 km, or 3.8×10^8 m. It travels around the Earth once a month. The distance it travels is the circumference of its orbit, $C = 2\pi r$, or about 2.4×10^9 m per month. Now, let's see, in a month there are . . . duh . . . 30 days (sometimes), 24 hours in a day, 60 minutes in an hour . . . so a month has $30 \times 24 \times 60 \times 60 = 2.6 \times 10^6$ seconds. Dividing 2.4×10^9 m by this time gives the Moon's velocity, $v = 9.3 \times 10^2$ m/s. To get the centripetal acceleration (6) tells us to square v and divide by the distance to the Moon. The result is

$$a_{cent} \ (Moon) = 2.3 \times 10^{-3} \ m/s^2. \quad (7)$$

So what? I'll show you so what. Let's work out the acceleration of the Earth's gravity to the distance

DEMO 2

(continued)

of the Moon. From (3) the acceleration of gravity at the distance to the Moon is

$$a_{grav}(Moon) = Gm_{Earth}/d^2_{Moon} \qquad (8)$$

where d_{Moon} is the distance to the Moon. Now, Newton didn't know G, but he could measure g, the acceleration of gravity on the Earth. Equation (4) gives us g. So let's play a trick. If we divide (8) by (4) we get

$$a_{grav}(Moon)/g = (r_{Earth}/d_{Moon})^2$$

or, multiplying by g,

$$a_{grav}(Moon) = g \times (r_{Earth}/d_{Moon})^2. \qquad (9)$$

Now we have a_{grav} in terms of things Newton could measure: g, the size of the Earth and the distance to the Moon. The radius of the Earth, $r_{Earth} = 6.4 \times 10^6$ m. The distance to the Moon, $d_{Moon} = 3.8 \times 10^8$ m. Plugging these numbers into (9) with $g = 9.8$ m/s^2, you get, voilà,

$$a_{grav}(Moon) = 2.8 \times 10^{-3} \text{ m/s}^2. \qquad (10)$$

Well, it didn't work exactly for Newton either. But notice this last number is pretty close to what we just got for the centripetal acceleration of the Moon in (7). It was just this "agreement" that suggested to Newton that the centripetal force holding the Moon in orbit *was* gravity and that gravity obeyed square of the distance to the Moon, as in (1) or (9).

We can play one more trick. Since for moons— or planets in orbit around the Sun—we have de-

DEMO 2

(continued)

cided the gravitational force *is* the centripetal force, let's equate them:

$$m_p v^2/r = Gm_p m_s/r^2. \qquad (11)$$

Now m_p will be the mass of a planet, m_s the mass of the Sun, and r the distance between Sun and planet. Once again the m_p's cancel. Multiplying both sides by r gives

$$v^2 = Gm_s/r. \qquad (12)$$

But the velocity of a planet is just the circumference of its orbit, C, divided by the time it takes to go around. That time is called the period, P. For circular orbits, the circumference is $C = 2\pi r$. Dividing C by P gives $v = 2\pi r/P$ and $v^2 = (2\pi)^2 r^2/P^2$. Plugging v^2 into (12) yields

$$(2\pi)^2 r^2/P^2 = Gm_s/r.$$

or, cross-multiplying,

$$r^3 = [Gm_s/(2\pi)^2] \ P^2.$$

All the junk in the brackets is a constant (a fixed number), so

$$r^3 = \text{constant} \times P^2 \qquad (13)$$

which is just Kepler's third law: the cube of a planet's distance from the sun is proportional to the square of the period. thus, the law of gravity can be used to derive Kepler's laws! (we have done this for circular orbits, but it also works for ellipses and you can get the other laws too.)

DEMO 3

Perhaps the easiest way to understand Equation (5) on page 38 is to ask: What is the acceleration of an object rotating in a circle of radius r at a velocity v?

By definition, acceleration is velocity divided by time. Time is distance divided by velocity. The only distance we are given is the radius, r. Therefore, the only time we can think of is $t = r/v$. Consequently, the acceleration $a = v/t$ must equal $v/(r/v)$. In other words, $a = v^2/r$ and the accompanying force is $F = ma$ or $F = mv^2/r$, which is Equation (5).

Next, focus attention on the first law, which says that the velocity of an object remains unchanged unless acted on by an outside force. Imagine a turtle moving at velocity v. We define the **momentum** of an object (the turtle) to be its mass times velocity (mv). If the turtle's mass remains unchanged, the first law then states that its momentum remains constant unless acted upon by an outside force. This is our first example of a **conservation law**, a special case of the

> **Conservation of Momentum:** The total momentum of a system does not change unless it is acted upon by an outside force.

While we are speaking of momentum, imagine a turtle in orbit around the Sun. It has an **angular momentum,** which is defined as its ordinary momentum times the radius of its orbit, or mvr. It turns out that angular momentum is also conserved. Thus, our **second law of nature**:

DEMO 4

Conservation of Momentum

To see conservation of momentum in action, get a basketball and a tennis ball. Holding the tennis ball on top of the basketball, drop them simultaneously from a height of about 1 m. After they bounce, you'll see the tennis ball shoot high into the air. It's pretty spectacular. But why does this happen?

Both balls get to the ground with nearly the same velocity v, but the basketball is a little ahead of the tennis ball. The basketball bounces and reverses direction. For a perfect ball, it still has velocity v (though in the opposite direction). Its momentum is Mv, where M is its mass. The basketball then clobbers the tennis ball, which is still moving downward with velocity v. The tennis ball's momentum is mv. Because the basketball is so much more massive than the tennis ball (M much larger than m), the basketball has a lot more momentum than the tennis ball. The basketball's momentum gets transferred to the tennis ball when they collide. You can guess what happens: the tennis ball really takes off.

Specifically, conservation of momentum tells us that

momentum (basketball + tennis ball)$_{\text{before collision}}$

= momentum (basketball + tennis ball)$_{\text{after collision}}$.

With this law plus conservation of energy (Chapter 3) you can calculate that under ideal conditions the tennis ball should go 9 times higher than the height from which you dropped it!

Conservation of Angular Momentum: The total angular momentum of a system does not change unless acted upon by an outside force.

Since conservation of angular momentum requires that mvr for the turtle remains constant, if for some reason it moves to a smaller orbit (smaller r) then it must speed up (larger v). You have seen conservation of angular momentum in action when a ballerina or an ice-skating turtle pulls in her arms and speeds up so fast she gets sick.

As engraved in the introduction, the laws of nature are the Ten Commandments of physics. They cannot be broken; they are invincible. In a real sense all physics is a search for the basic laws and an application of them.

Sir Isaac Newton— The inventor of mechanics, Newton formulated the laws of motion that lie at the heart of almost every physics problem.

Finally, Newton's second law had enormous repercussions for not only physics, but philosophy as well. First consider a car traveling with velocity 100 kilometers per hour in New York City. The driver suddenly decides he can't take it anymore, and accelerates out of town with an acceleration, a. From Equations (5) and (7) in the box on pages 30–31, it is then easy to compute what the velocity will be after an hour, as well as the distance from New York.

Similarly, say from the law of gravitation, we know the gravitational force, F_{grav}, on a planet of mass m. Newton's second law, $F = ma$, then gives us its acceleration: $a = F_{grav}/m$. As we did for the car, we can then compute the planet's velocity and position after any time. Indeed, this is basically how Newton showed the planetary orbits were ellipses.

ESOTERIC TERMS
(es-ə-'ter-ik tərms)
(for the initiated)

Physics is not poetry, though it is poetical. The following terms have one and only one meaning in physics. Thou Shalt Use Them Properly.

- *Velocity*—The change of distance with time. Velocity and speed are often used interchangeably, even among physicists, but technically velocity refers to a speed *in a given direction.* That is, we may say the speed of a car is 100 kilometers per hour, but its velocity is 100 kilometers per hour north. **Thus the velocity of an object is altered if its direction of motion changes.** A ball whirled around on a string has constant speed but changing velocity. Since velocity is the change in distance per unit time, it is measured in meters per second, or m/s (q.v. *Wayne's World* re: "velocity of hurl").

- *Acceleration*—The change of velocity with time. Because the velocity of the above ball is changing, it is accelerating, even though its speed is constant. If acceleration is the change in velocity per second, and velocity is measured in meters per second, then acceleration must be measured in meters-per-second per second, also written as meters per second squared, or m/s^2. Do you get-it get it?

ESOTERIC TERMS
(*continued*)

- *Force*—A push or a pull, nothing more meta-physical. The precise definition is given by the second law, $F = ma$. A 1-kilogram mass accelerating at 1 meter per second^2 is said to experience a force of 1 newton (N). To elaborate, we apply a spring to the standard kilogram mass, housed in Paris (ah, to be an inert standard setter in Paris). If the mass accelerates at the rate of 1 m/s^2, we *define* the force applied to be 1 N. If it accelerates at 2 m/s^2, we say the force is 2 N, etc.

- *Momentum*—The product of an object's mass and velocity *(mv)*.

- *Mass*—This term is not defined in physics but can be thought of as the amount of "stuff" in an object. One can also view it as an object's "resistance to acceleration," found by inverting the second law: $m = F/a$. That is, we find a 1-newton spring as above. We apply it to an unknown mass and discover that it accelerates at 2 m/s^2. That means it has a mass of ½ kg: $m = F/a = 1$ N $(2$ m/s$^2) = $ ½ kg).

- *Inertia*—An object's resistance to acceleration.

We see that, given the initial position and velocity of an object and the forces acting on it, we can predict the behavior of that object for any time in the future (or past). We say Newton's laws are *deterministic*, or contain "zero defects." More generally, given the initial position and velocity of *all* the particles in the universe and the forces acting between them, the second law determines the entire future history of the universe!

For this reason, determinism, mechanism, and the metaphor of the universe as a giant clockwork came into prominence in the seventeenth century. According to Enlightenment philosophers, God was needed now only to start things off (carefully choosing the initial position and velocities of all the particles) and Newton's laws would do the rest. This was one of the great chapters in the ongoing debate about the relationship between science and religion. The clockwork mechanism of the Newtonian universe has recently required repair, but we will leave that for Chapter 10.

SUMMARY

 Key word: Force.

 Key definition: $F = ma$.

 Key 'cept: Practically all of physics hinges on $F = ma$, which could be introduced only by first overthrowing Aristotelian philosophy.

CHAPTER 2

THE ELUSIVE ATOM

I'VE GOT GAS

YOU MUST REMEMBER THIS

Atoms cannot be seen. To show that the world was made of particles a million times smaller than objects visible to the naked eye was so difficult that their existence was not established beyond reasonable doubt until the end of the nineteenth century.

Aristotle's Atomic Mess

Just as the establishment of mechanics was largely the freeing of scientists' imagination from the Aristotelian concept of motion, the establishment of chemistry—and with it atomic physics—was largely a reaction against Aristotle's notion of substance.

In the Prologue we saw that the pre-Socratics Leucippus (if he existed) and Democritus arrived at the idea of atoms in order to explain the phenomenon of change. The word "atom" means indivisible, and atoms were to be the basic building blocks of nature. Nevertheless, you must realize that to the Greeks and their successors, the concept of atom was an abstract **hypothesis** that might explain certain phenomena. But at the time there could be no experimental evidence to support it. And there would be no evidence for a thousand years.

If this seems like a strange statement in this age when we take atoms for granted, ask yourself: how do you know atoms exist? Yes, you. Probably you read about them in grade school. But what experiment can you propose to give evidence of their existence? Your Ginsu knife may slice and dice, but it's not going to cut a brick into atoms. To most people there's little difference between atoms and magic.

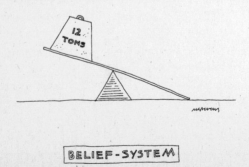

BELIEF-SYSTEM

To illuminate the difficulties more sharply, pretend you are a scientist in the sixteenth or seventeenth century. The villain Aristotle had rejected atoms and instead embraced Empedocles's notion of four elements—earth, air, fire, and water. All remaining substances on the Earth arose from combinations of these elements. As a follower of Aristotle, it is difficult for you to conceive of anything more fundamental. But the result, for scientists trying to prove this theory, was centuries of blood, sweat, and tears. Thanks, Ari.

Moreover, the Master held that once a new substance is formed, all traces of its ingredients are erased. To take a modern example, we know that limestone (chalk to laycreatures) is composed of calcium, carbon, and oxygen. But as an Aristotelian, you scoff at this. Limestone is an entirely different substance; once it has formed, the calcium, carbon, and oxygen vanish. Along with the fact that everything is ultimately made of earth, air, fire, and water, you are convinced that it is a waste of time to search for more basic units of construction.

The next great obstacle to an understanding of atoms is that you lack the modern concept of a chemical element—a basic substance that cannot be transformed by any chemical reaction into another substance. Following the Aristotelian quest for perfection, you have attempted to make the most perfect metal—gold—in your alchemy-laboratory. This seems a perfectly reasonable endeavor. During your years in the laboratory you added zinc to copper, which is yellowish, and thereby transmuted copper into the more yellow brass. Why cannot brass be transmuted into something even more yellow—gold? Without the idea of an element, there is no reason to think this can't be done.

The chemical situation at the start of the seventeenth century was even worse than this. Chemical ideas were mixed with astrological ideas (the behavior of iron was governed by the red planet Mars; followers of Jean Dixon have evidently stopped thinking in the seventeenth century). And because Aristotle taught that things follow their nature, some chemicals reacted because they felt "sympathy" for each other, and others refused to react because they "abhorred" each other. Indeed, by the seventeenth century, chemistry had become the Yugoslavia of science. How could such a mess be sorted out? The answer lay in atoms, but to prove it took another three hundred years of false starts, missed opportunities, and inspired guesses.

WHO'S WHO ☞

Evangelista Torricelli (1608–1647)
Discovered air pressure, maybe invented barometer, caused scandal, helped glass tube industry.

Robert Boyle (1627–1691)
The first rational chemist. Codified laws for air pressure. Believed in atoms and in the quite modern idea of nature being comprised of elements. Admitted mistakes.

Daniel Bernoulli (1700–1782)
Derived Boyle's law from assumption that atoms exist. Ahead of his time. Was ignored.

Mikhail Lomonosov (1711–1765)
Wrote poetry. Also first to articulate conservation of matter. The Russian Ben Franklin.

Joseph Black (1728–1799)
A nice guy. Isolated carbon dioxide ("Dioxide, go to your cell and stay there!"). Also proposed law of specific heats.

**WHO'S
H
O**
☞

(continued)

Antoine Lavoisier (1743–1794)
Showed that phlogiston theory is just a lot of
hot air, and explained combustion, "found"
modern chemistry, lost his head in French Revolution. (The rewards for scientific discovery
have since improved.)

John Dalton (1766–1844)
A dull man, but introduced modern concept
of atom. Launched the atomic weight guess-
ing game.

Amedeo Avogadro (1776–1856)
His real name was longer. Introduced idea of
molecules. Was ignored, but the number
named after him survives.

Dmitri Mendeleev (1834–1907)
Survived childhood, independently invented
periodic table. Started wave of periodic table-
settings.

Albert Einstein (1879–1955)
Explained random motion of pollen in water
by assuming existence of molecules. See Chapter 5 for his accomplishments in the field of
relativity.

TOTALLY TUBULAR TORRICELLI

The attack on Aristotelian notions began in the fifteenth
and sixteenth centuries with the publication of Lucreti-
us's *De rerum natura* and Heron's *Pneumatica*, both long

out of print. The former was a poem written in the first century B.C. expounding the atomic hypothesis. (Science in those days was considered a fit subject for poetry.) In the latter, Heron of Alexandria (fl. A.D. 60) described pneumatic devices to open temple doors and to deliver holy water when a coin was deposited (the first slot machine). He viewed air as made of tiny particles, and his ideas are reflected in the outlook of natural philosophers such as Galileo and Robert Boyle.

The properties of air turned out to be crucial in the hunt for atoms. A large step in dissecting air was taken by a variety of Italian scientists, culminating in the efforts of Galileo's assistant Evangelista Torricelli. As textbooks tell the tale, Torricelli poured various liquids—seawater, wine, mercury—into long glass tubes sealed at one end. He inverted the tubes, standing them upright in pans filled with the same liquids, and lashed them to the masts of ships. The fluids fell considerably from the top of the tubes, leaving a gap. Torricelli then observed that

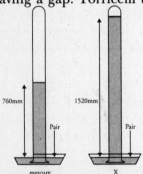

Torricelli's Experiment—
Whoever did it, the experiment showed that the vacuum at the top of the tubes is not sucking up the liquid below. Rather, the atmosphere exerts a fixed pressure (P) on a pan of fluid. This pressure is sufficient to support a mercury column 760 mm high (left). If a column of mystery fluid (right) were found to rise to 1520 mm, we would conclude it has half the weight, and therefore half the density, as mercury.

the lighter liquids formed higher columns than the heavier ones. Indeed, if wine were, say, twice as light as water, then the wine column would be twice as high as the water column. A column of mercury, the densest fluid known, rose about 760 millimeters (mm).

Torricelli's (Viviani's—whoever's) experiments created a sensation. Why, for God's sake? Remember your Aristotle: nature abhors a vacuum. Torricelli filled his tubes completely before inverting them. So when the liquid settled down, the gap at the top of the tube must contain nothing—a vacuum. The public is astounded. The tabloids go wild.

NATURE ALLOWS VACUUM.
ARISTOTLE REFUSES TO ADMIT DEFEAT.
SAYS, "IT AIN'T SO."

"Everyone knows," claims Greek, "that you can only drink soda because when you try to create a vacuum in your mouth by sucking on a straw, the soda rushes in."
Philosophers Weigh Results

Weight. Hmmm. From his investigations Torricelli concluded that no, a vacuum at the top of the tubes was *not* sucking up the fluid. Rather, the *atmosphere* was pressing down on the liquid in the pans and supporting the column. The heavier the liquid, the less the atmosphere could support.

We live submerged at the bottom of an ocean of elemental air, which by unquestioned experiments is known to have weight.

And so Torricelli established the notion of atmospheric pressure: it is the pressure sufficient to support the weight of a column of mercury 760 millimeters in height.

Apart from creating a scandal, he (Berti—whoever) incidentally invented the barometer.

For those of you who insist Aristotle was right, within a few years Torricelli's hypothesis was confirmed by several others, including Blaise Pascal and Otto von Guericke (1602–1686), who invented the first air pump. Placing an entire barometer in a sealed chamber, von Guericke confirmed that as he pumped the air out of the chamber the barometer's mercury fell. This is as it should be, since there is no longer any atmospheric pressure to hold it up.

PUMP UP THE VOLUME

Robert Boyle— He had faith in atoms and elements when few others believed. He is considered to be the first rational chemist.

Not long after, von Guericke's work came to the attention of Robert Boyle, who promptly ordered his own pump built to continue the investigations. Boyle is considered by many to be the first rational chemist (and one of the first to acknowledge experimental errors) and he played a role analogous to Galileo's in the development of mechanics.[1] He published his air-pump results in 1660 in *New Experiments Physicomechanical, Touching the Spring in the Air and Its Effects, Made for the Most Part, in a New Pneumatic Engine.*

This treatise contained his verification of Torricelli's hypothesis. It also contained Boyle's Law. One of the experiments Boyle reported consisted of a glass U-tube

[1]Mary Boas Hall, "Robert Boyle," *Scientific American* (Aug. 1967, p. 96).

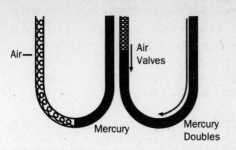

Air—
Air Valves
Mercury
Mercury Doubles

Boyle's Law— As the amount of mercury increases, the volume of air decreases proportionally. Air compacts like a spring. (We have ignored the added pressure of the air itself.)

filled with air in one arm and mercury in the other. As he added more mercury to the one arm, the volume of air in the other decreased.

Indeed, as he doubled the amount of mercury, the volume of air halved, and so on. In doubling the amount of mercury, Boyle doubled the pressure exerted by the mercury on the air—and doubled the pressure the air exerted on the mercury. Air was behaving like a spring. Indeed, these observations indicated that when the temperature is fixed, the volume V of a gas is inversely proportional to its pressure P:

$$V = \text{constant}/P,$$

or

$$PV = \text{constant}.$$

This equation is known as Boyle's law. A century later, in 1787, Jacques Charles showed that if the pressure were kept fixed, then as the gas was heated, the volume would increase in proportion to the temperature, T. That is, $V = constant \times T$. Combining this with Boyle's law we get:

$$PV = \text{constant} \times T,$$

which is a form of the famous **ideal gas law** taught to all high-school chemistry students. This law does just what the equation says: it describes the relationship among P, V, and T for simple gases.

By the way, an important moral to take from all these post-air-pump results, as with Galileo's application of the telescope, is contained in what we might call the

> **Principle of Magnification:** New discoveries follow
> on the heels of new equipment.

Now, you might wonder what this has to do with atoms. Well, it turns out that in 1738, even before the experiments of Charles, Daniel Bernoulli showed that one could derive Boyle's law if one assumed that gas pressure was produced by invisible particles colliding with the container wall, particles that obeyed Newton's laws! Bernoulli's proof is, apart from being brilliant, fairly simple, and we reproduce it in Chapter 3. But it was also far ahead of its time and evidently ignored. Bernoulli's failure to convince contemporaries of the reality of atoms leads us to another observation:

> **The Graveyard Principle:** To be behind one's time
> is permanent death. To be ahead of one's time may
> be temporary death. But Confucius say: dead is dead.

If Bernoulli's ideas had not been interred, perhaps the idea of atoms would have been accepted by the eighteenth century. But they were and it wasn't. So much for progress.

BEYOND THE BOYLING POINT

To return to the stumbling progress of seventeenth-century chemistry, the atomic hypothesis still had no real evidence to support it. But it had adherents. Its chief advocate was, again, Robert Boyle. In 1661, shortly after announcing his gas laws, he published a Galilean dialogue, *The Sceptical Chymist.* Influenced by Heron and Enlightenment philosophy, Boyle's self-proclaimed goal was to put chemistry on a mechanical footing and consider it part of "a great piece of clockwork." To achieve this end, he vigorously attacked such alchemical notions as "sympathy" and "abhorrence," and tried to show there was no evidence for Aristotle's four elements. Indeed, Boyle introduced an "almost modern"[2] definition of an element, that is, *a chemical substance that cannot be broken down into any other.*

The Sceptical Chymist was enormously influential and convinced many chemists to look for a particulate explanation of reactions. On the other hand, the road to atoms took a major detour with the introduction of phlogiston by Georg Ernst Stahl (1660–1734).

When wood is burned, only ashes remain, and the weight of the ashes is less than the original weight of the wood. Stahl proposed that the difference was carried off by a mysterious substance he named "phlogiston" (from the Greek word for "flammable"). Unfortunately, it was also known that when some metals were heated in air, they gained weight. The observation did not deter Stahl; in this case the phlogiston was said to have negative weight. Despite the fact that the properties of phlogiston were impossible to pin down (and you shouldn't try), the theory was accepted by most chemists of the eighteenth century. This prevented a clear understanding of combustion, which turned out to be the key to everything.

The problem was compounded by the fact that at the time no one realized that air was composed of more than one substance. Air was air. (Again, you should quiz yourself on how you might show air has many components.) Consistent with the Magnification Principle, the breakthroughs began with the invention of the pneumatic trough by Stephen Hales (1677–1761). This was a simple device for collecting gases above water that were given off by substances burning in a furnace.

---

[2]Not quite modern because he believed in the transmutation of elements.

With Hales's invention, Joseph Black isolated a gas that was given off by heating magnesium carbonate, which he was investigating as a "lithontriptic" substance (look it up, impress your friends). Black recognized the new gas was heavier than ordinary air. Heating limestone and quicklime produced the same stuff. Black called it "fixed air" (because it was "fixed" in lime). Today we call it carbon dioxide (CO_2).

Moreover, after heating the limestone ($CaCO_3$), Black could take the residue (quicklime—CaO) and dissolve it in water to make limewater. If he then bubbled carbon dioxide through the limewater, the original limestone would reappear—and it would be the same amount he originally started with. This showed for the first time that a gas (CO_2) could react with a substance (CaO) to produce a totally new substance—limestone ($CaCO_3$). The discovery was a blow against the Aristotelian decree that once a substance was formed its ingredients disappeared.

Once Black isolated carbon dioxide, work on gases proceeded quickly. Hydrogen had already been isolated by the Russian Mikhail Lomonosov (see feature below). He found that when he treated any nonprecious metal with acid, an "inflammable vapor" was given off that "is nothing else but phlogiston." Lomonosov's work remained unknown in the West when in 1766 hydrogen was again identified as a distinct substance by the English scientist Henry Cavendish (1731–1810). Within a few years Carl Wilhelm Scheele and Joseph Priestley (1733–1804) independently isolated a gas that was "purer" than ordinary air. In "pure air" candles burned more brightly than in common air and animals lived longer. Priestley termed it "dephlogisticated air." Lavoisier would soon name it oxygen. At about the same time Daniel Rutherford isolated nitrogen, which he called "mephitic air"—atmospheric air saturated with phlogiston.

At this juncture several clues emerged on the road

MIKHAIL LOMONOSOV (1711–1765) . . . .

Mikhail Lomonosov was born the son of a fisherman near the White Sea. At the age of 19 he disguised himself as the son of a priest to get into St. Petersburg's Slavic-Greek-Latin Academy. Someone smelled something fishy, and the ruse was exposed, but because of his exceptional brilliance, Mikhail was allowed to remain. After study in Germany, he returned to Russia where he quickly established himself as the Russian Ben Franklin. He wrote poetry, became famous as the author of the first Russian grammar and performed experiments in chemistry and electricity. His investigations led him to attack the phlogiston theory and he became the first to articulate the conservation of matter. There is evidence that Lavoisier was aware of but did not acknowledge his work. ("Either you do the calculation or you get the credit.") Lomonosov also helped found Moscow State University, which is now named after him. The poet Pushkin said of Lomonosov, "He founded the first Russian University. One might better say he was our first university."

to atoms. Cavendish discovered that "inflammable air" (hydrogen) was about 10 times lighter than common air (the true answer is 14.4). He also carried out experiments that suggested that two volumes of hydrogen burned in the presence of one volume of oxygen would produce water vapor. As we will soon see, this result is easily explained in terms of atoms, but Cavendish was a dedicated phlogistonist, and interpreted the reaction incorrectly.

LAVOISIER: *LA VOILÀ!*

In a word, phlogiston was the chemical equivalent of Bob Dole—always there when you don't want him. The

*Antoine
Lavoisier—
The "founder" of
modern chemistry,
his experiments
sent the false
theory of phlog-
iston up in flames.*

honor of ridding the world of phlogiston, and by the way creating modern
chemistry, is usually accorded to Antoine Lavoisier. By burning various substances under controlled conditions
and keeping strict track of the prod-
ucts, he determined that a burning substance does not give off "phlogiston,"
but rather combines with oxygen.
When wood is burned, the carbon combines with oxygen to form carbon dioxide, which usually disappears up the
chimney. Hence the weight of the ash
is less than the weight of the wood.
When iron burns, it combines with oxygen to form iron oxide (rust). Thus the
weight of the iron increases, and the surrounding air is
deprived of a little oxygen.

In his controlled experiments Lavoisier found that the
total weight of everything (wood + air) before burning
equaled the total weight of everything after burning (ashes
+ smoke + air). All of his results were published in 1789 in
his famous *Traité Elementairé de Chimie*, which was to chemistry what Newton's *Principia* was to physics. In it he assumed the law that Lomonosov had posited years earlier:

> **Conservation of Matter:** Matter is neither created
> nor destroyed; it is merely changed in form.

This is our **third great conservation law** and thou shalt
obey it. The conservation of matter in chemical reactions
is easily explained by the existence of atoms: the total
weight of the chemicals does not change because the
total number of atoms in the reactions does not change.
However, this was not realized at the time.

ANTOINE LAVOISIER (1743–1794)

Antoine Lavoisier was born in Paris; his father was a Parliamentary prosecutor and his mother a wealthy advocate's daughter. Not surprisingly, he received the finest education available at the time and took a bachelor's degree in law, though he also pursued mathematics, astronomy, anatomy, physiology, meteorology, and chemistry. His first paper on chemistry appeared at the age of 22. By the time he was 25 he was elected to the French Academy of Sciences.

Lavoisier is credited with putting chemistry on a modern footing. By showing that in combustion oxygen combines with the substance being burned, he disproved the phlogiston theory and paved the way for a correct analysis of chemical compounds. Unfortunately, he was also a member of the *Ferme Générale*, a corporation under royal charter to collect taxes. During the French Reign of Terror, members of the Ferme were arrested as having oppressed the people and robbed the public treasury, a taxpayer reaction that the IRS might find instructive. Lavoisier was guillotined in 1794.[1]

---

[1]For more on the chemical revolution, see John Maxson Stillman, *The Story of Alchemy and Early Chemistry* (New York: Dover, 1960)

Once Lavoisier established modern chemistry, the pace of events accelerated (note use of physics term in everyday language). One raging debate of the age was whether compounds had variable or fixed proportions of their ingredients. That is, you can vary the amount of cheese in an omelette, but it remains an omelette. Similarly, you can vary the amount of zinc and copper in brass, but the result is still brass. Now, what about limestone? Can you change the proportions of calcium, carbon, and oxygen and still make chalk?

The issue was largely decided in the late eighteenth century when Joseph Louis Proust (1754–1826) engaged in the preparation of copper carbonate (copper + carbon + oxygen) and other compounds. No matter how much of the ingredients he started with, the proportions consumed in the reaction remained the same, say always 5.25 times as much copper as carbon, by weight. This was great evidence against the omelette theory. That is, chemical compounds always have fixed proportions of their ingredients. For this reason Proust's law is called the **law of constant proportions**. (That other Proust, the French author, discovered the as-yet-unrecognized law of disproportion: the total of a cookie dipped in tea caused a chain reaction that led to a three-volume memoir.)

In retrospect, we can see that Proust achieved his results because the same number of copper atoms always combine with the same number of carbon and oxygen atoms (copper carbonate: $CuCO_3$). Indeed, chemists at the time used Proust's work to argue for the existence of atoms.

ELEMENTAL, MY DEAR DALTON

One of these chemists was John Dalton. Dalton was not the first to posit the atom but he was the first to give it more-or-less modern properties. Dalton held that all elements were made of small, indivisible particles. Atoms of a given element were identical but differed from atoms of all other elements. Chemical combinations took place when atoms from different elements joined each other. Knowing that the popularity of a theory is enhanced by publicity, Dalton showed himself to be a good publicist by calling his particles "atoms."

Dalton's hypothesis led to two promising ideas. The first is the concept of **atomic weights**, which means exactly what it says: the atoms of different elements have different relative weights. Now, atoms are rather small and you can't weigh one at a time; you must weigh large volumes. This Dalton could do. For example, his data showed that 85⅔ parts oxygen by weight combined with 14⅓ parts hydrogen formed water. The trouble was, he didn't know the chemical formula for water. He knew only that water was hydrogen and oxygen, not necessarily H_2O. In fact, Dalton's rule was to choose the simplest possibility: Oxygen was *O*, hydrogen was *H*, and water was *HO*. In that case, if the weight of a hydrogen atom is chosen as 1, the weight of an oxygen atom is then 85⅔ divided by 14⅓ or nearly 6 (the wrong answer). Nevertheless, in this manner he was able to compile a table of (incorrect) atomic weights.

Closely related was his **law of multiple proportions**. Some elements were known to form more than one compound. Carbon combines with oxygen to form carbon monoxide (CO) as well as carbon dioxide (CO_2). Notice that for a given amount of carbon, say 10 grams, carbon dioxide contains twice the weight of oxygen as carbon monoxide. Although Dalton lacked modern chemical formulas, he did notice that all the compounds he investigated behaved like carbon monoxide and carbon dioxide. That is, a fixed weight of element A always combined with either one or two portions of element B, never with a fraction. With our modern notation, it is clear that this is because twice as many oxygen atoms attach themselves to carbon in carbon dioxide as in carbon monoxide. If you should ever doubt the importance of subtle atomic variations in gaseous compounds, just compare the smell of your own breath with the rush hour fumes on L.A.'s Ventura Freeway.

Unfortunately, Dalton's ideas about keeping atoms simple tripped him up in the end. The experiments of Joseph Gay-Lussac (1778–1850) with gases were causing him indigestion. Gay-Lussac found, for instance, that one volume of nitrogen combines with one volume of oxygen to make one volume of nitric oxide (NO). This clearly suggested that equal volumes of gases contain equal number of atoms. Dalton couldn't buy it. He knew, for example, that water vapor (steam) was *lighter* than oxygen. But how can this be? Steam consists of hydrogen *plus* oxygen, so one steam particle should weigh *more* than one oxygen atom. So if, as Gay-Lussac said, one liter of steam held the same number of particles as one liter of oxygen, the liter of steam should be *heavier* than the liter of oxygen, not lighter.

There was a second problem. Gay-Lussac showed that

2 volumes hydrogen + 1 volume oxygen → 2 volumes steam.

Now this was ridiculous. If one volume of oxygen (O) contains n atoms, then surely it can only combine with n atoms of hydrogen (H) to make n atoms of steam (HO, by Dalton's rule). So you should get only one volume of steam out, with hydrogen left over. How to resolve these paradoxes? Sulk. Pace. Scratch one's head.

INTRODUCING THE MOLECULE

In retrospect we see that Dalton lacked the concept of **molecules**. To him oxygen was O and hydrogen H. To resolve the paradoxes, we need something else. The prize went to Amedeo Avogadro, who introduced the concept of molecules and was ignored. Accepting Gay-Lussac's results, he postulated:

1) Equal volumes of gases contain equal number of particles.

2) The particles can either be single atoms or molecules, depending on the gas. A molecule is the smallest collection of atoms of a given substance that occurs naturally.

For example, oxygen and hydrogen are found in nature

not as O and H but as **diatomic** molecules O_2 and H_2. Using Gay-Lussac's observations above, we thus postulate:

$$2H_2 + O_2 \rightarrow 2H_2O.$$

Here we have used conservation of matter: the number of H's and O's on the right must equal those on the left; otherwise atoms would be disappearing. If you have been through high-school chemistry, you know this painful procedure of accounting as *balancing* the equation.

Notice how Avogadro's hypothesis incorporates Gay-Lussac's results. If equal volumes of gases contain equal number of molecules, then two volumes of H_2 plus one volume of O_2 indeed produce two volumes of steam (H_2O), as observed. Moreover, this resolves Dalton's density paradox. A molecule of steam (H_2O; atomic weight 10) is indeed lighter than a molecule of oxygen (O_2; atomic weight 16), so a volume of steam weighs less than a volume of oxygen.

AVOGADRO—THE ANSWER

It would take another fifty years for chemists to realize Avogadro was right, but they finally woke up. Only at that point could one correctly calculate atomic weights (since you need to know the correct number of atoms in a volume of gas). That led to the development of the **periodic table**. Several scientists noticed that if you arranged the elements in order of increasing atomic weight, their chemical properties repeated periodically. For instance, helium (atomic weight 4) and neon (atomic weight 10) didn't react with anything. This suggested placing neon and helium in the same column of a table. Credit for the periodic table usually goes to Dmitri Mendeleev, though as usual, credit is in the eyes of the nationality. One Lothar Meyer appears to have done it

ESOTERIC TERMS
(es-ə-'ter-ik tərms)

- *Confusion*—The state of chemistry for three hundred years.

- *Element*—A basic chemical substance that cannot be transformed into another substance by chemical reactions.

- *Atom*—The smallest unit of a chemical element. There are about 100 known elements, of which about 90 are naturally occurring.

- *Molecule*—The smallest collection of atoms of a substance that occurs naturally. For instance a water molecule consists of two hydrogen atoms and an oxygen atom: H_2O.

- *Pressure*—The force per unit area exerted by a gas or fluid on something else. Atmospheric pressure is 100,000 newtons per square meter. Puckish scientists call illustrations of this "figs Newton."

- *Boyle's law*—The statement that for a gas at constant temperature, PV = constant, where P is the pressure and V is the volume.

- *Atomic weight*—Loosely, the weight of an atom of a given element, with hydrogen taken to be 1.

- *Avogadro's number*—The winner of last week's lottery. Also, the number of molecules of a gas in 22.4 liters. Equal to 6×10^{23}.

independently at the same time. In any case, Mendeleev's table held gaps, but by the position of the gaps he was able to predict the existence of new elements, which were subsequently discovered.

Despite all this, resistance to the idea of atoms continued. (Remember, no theory is fully accepted until the last of its opponents dies off.) The mop up operation was left to Albert Einstein. In 1827 a botanist named Robert Brown had observed that microscopic plant pollen suspended in a drop of water continually jiggle around for no apparent reason. Only in 1905 did Einstein show that molecules colliding with the spores would produce Brownian motion, as it was called. Einstein's calculation also provided the best value of "Avogadro's number"—the exact number of molecules that occupy a standard volume of gas. Every chemistry student is forced to memorize this figure: 6×10^{23} molecules per 22.4 liters.

With the calculation of Avogadro's number, atoms finally became real. A few crotchety old men objected even then, but when they saw flashes of light produced by the radioactive decay of atoms, even they capitulated. However, that story belongs to nuclear physics and we defer it to Chapter 6.

SUMMARY

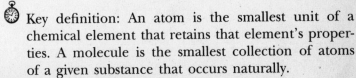

Key word: Atom.

Key definition: An atom is the smallest unit of a chemical element that retains that element's properties. A molecule is the smallest collection of atoms of a given substance that occurs naturally.

Key 'cept: Conservation of matter; nothing disappears in chemical reactions.

CHAPTER 3

THE ENERGY CRISIS

SCIENTISTS IN HEAT

YOU MUST REMEMBER THIS

Like atoms, heat is so intangible that it was one of the last concepts in classical physics to be sorted out. In the process, the science of thermodynamics was created. Pollyannas who believe anything is possible should be subjected to a course in thermodynamics.

CALORIE COUNTING

In the same book in which Lavoisier destroyed phlogiston, an erroneous concept that plagued scientists throughout much of the eighteenth century, he introduced **caloric**, an erroneous concept that plagued scientists for much of the nineteenth century. His fame does not seem to have suffered. Remember, one only arrives at the right answer after making all possible mistakes and, anyway, one should judge people by their best work, not their worst.

Caloric is a bit easier to understand than phlogiston. You probably worry so much each day about calories that you overeat. There is an oven with the brand name Caloric. This time we must blame the Romans: "caloric" from the Latin *calor*, meaning "heat." It was not news, even in the eighteenth century, that burning coal produces heat. Though Lavoisier understood that no mysterious fluid, phlogiston, combines with the coal in this process, he was still a child of his time. Heat itself must be a fluid, given off during combustion. Lavoisier was not the first person to have this idea, but he gave it a mass-market name, caloric. It stuck.

Indeed, the branch of physics that deals with heat is termed **thermodynamics**, which literally means "the dynamics of heat." It seems strange to us that something as intangible as heat should move, but in those days, before physicists were called physicists, they believed that heat was something like water.

If most natural philosophers of the time believed in caloric, a few did not. A severe blow against the idea was struck by Massachusetts-born Benjamin Thompson, who was a spy for the British in the days leading to the American Revolution. (Curiously, most physics texts fail to mention this fact.) On the verge of being tarred and feathered, he abruptly quit the colonies in 1775, returned to fight against them, was accused by the British of spying for the French, ended up in Bavaria and became Count Rumford, by which name he is revered. Improbably, he also married Lavoisier's widow.[1]

Apart from campaigning against drunkenness and being miserable with the widow Lavoisier, Count Rumford supervised the munition works in Munich, where he was in charge of cannon boring. He noticed that the process generated an immense amount of heat. The count reasoned that if heat were a fluid, a brass cannon could only hold so much, and sooner or later it should run out. But in a series of controlled experiments he found this wasn't the case. The amount of heat produced depended on the amount of time spent boring the cannon. With a dull drill bit, he could bore practically forever and extract no metal but practically an infinite amount of heat. If you have ever done some drilling in your basement shop with an old bit, you've noticed the same thing but haven't become famous. In this way Rumford came to the conclusion that heat was associated with mechanical work, not caloric. Indeed, Rumford speculated that heat had something to do with motion.

To appreciate the significance of Rumford's conclusion, we must first go back a few decades. In about 1760, the same Joseph Black who discovered carbon dioxide noticed that when he heated a block of ice, its temperature didn't rise until all the ice was melted. Some heat was going into melting the ice but not raising the temperature. From this Black observed that temperature and heat were two different things. He also observed that when he mixed mercury at, say, fifty degrees with an equal amount of water at, say, zero degrees, the final mixture would be at only about one degree. That is, the

---

[1]Mitchell Wilson, "Count Rumford," *Scientific American* (Oct. 1960, p. 158).

**WHO'S
WHO**☞

Antoine Lavoisier
We've talked enough about him already.

Benjamin Thompson, Count Rumford
(1753–1814)
British spy and Bavarian bigwig. Bored cannons stiff, proving heat was not caloric fluid. Continued to work with the body of research left behind by Lavoisier (and had a warm time with Lavoisier's widow, whom he eventually married).

Joseph Black (1728–1799)
Still a nice guy; introduced concept of specific heat, which is unrelated to that setting in a car's heater where the warm air just blows on your feet.

James Prescott Joule (1818–1889)
The Energizer, paddled his way into science history. Namesake of a unit of energy. Also showed heat was a form of energy.

William Thomson, Lord Kelvin (1824–1907)
Devised first statement of second law of thermodynamics, which makes him way cool.

Rudolf Clausius (1822–1888)
Gave thermodynamics its mathematical footing. Introduced entropy, made world miserable.

Ludwig Boltzmann (1844–1906)
Many contributions to thermodynamics. Tried to derive second law from Newtonian mechanics. Couldn't take it anymore. Can you blame him?

mercury would cool by forty-nine degrees but the water would only heat up by one degree. This might seem strange: If the water is absorbing the heat given off by the mercury, shouldn't the fall in temperature of the mercury equal the rise in temperature of the water?

No. Once again, Black realized heat and temperature must be different. In that case, how are they related? His observations led him to the concept of **specific heat**. Specific heat is defined as the amount of heat it takes to raise the temperature of one gram of a substance one degree Celsius. Specific heats vary widely from substance to substance. Water has a much higher specific heat than mercury, thus it takes a lot more heat to change the temperature of water by one degree than it does to change the temperature of mercury by one degree.

Indeed, the **calorie** is defined to be the amount of heat needed to change the temperature of one gram of water by one degree Celsius.[2] Usually specific heat is given the symbol C. Now, let Q be the amount of heat (in calories) you put into a block of a substance with a mass of m grams, and let ΔT be the temperature change. (Delta, the little triangle Δ, is the standard symbol for a change in something.) Then in symbols the relationship between the amount of heat and the temperature change will be:

$$\Delta T = Q/mC,$$

or

$$mC\, \Delta T = Q. \qquad (1)$$

The top version of the formula makes it clear that for a given amount of heat Q, and a given mass m, then the smaller C, the larger ΔT. This is exactly what Black

---

[2]To confuse things, the calories you sweat out every day in the gym are 1,000 of these calories; technically they are called kilocalories.

found. For instance, the specific heat of mercury is about .03, so it takes roughly 30 times *less* heat to change the temperature of a gram of mercury one degree than a gram of water.

JAMES JOULE: FOREBEAR OF THE ENERGIZER BUNNY

James Prescott Joule—
One of about five men who concluded independently that heat was a form of energy. But he got the credit—and the unit.

Well, if heat was not temperature or caloric, what was it?

The answer was largely provided by a series of experiments performed by James Prescott Joule in the 1840s. Joule built a contraption that consisted of some weights attached by strings and pulleys to a paddle wheel inside an insulated container of water. As the weights fell, the paddle wheel turned and by friction heated up the water. The temperature change, via the concept of specific heat, corresponded to a certain amount of heat put into the water. Joule discovered that a given amount of heat put into the water always corresponded to the same distance fallen by the weights.[3]

Now, what is changing as the weights fall? They are losing **potential energy**. We will make this statement more precise in a moment. But a better way of saying what Joule found is that a given amount of heat put into the water always corresponds to the same amount of potential energy lost by the weights. The conclusion? **Heat is a form of energy.**

---

[3]Actually, this experiment seems to have been performed a decade earlier by Sadi Carnot, who perished, not published.

Energy. Kids have a lot of it, the President wants to tax it, you may lack it in the morning, acupuncturists reverse the flow of it, there was a crisis of it in the 1970s, TV transcendentalists harness the psychic form of it. Energy. But what does energy *mean*? Well, judging from all the ways people use the term, it means absolutely nothing. Good, now we can dispense with the rest of the book. No. Rather, let us put you on the path to True Enlightenment. The true, precise, unique, and unalterable definition of energy is

the ability to do work.

Work. Something to be avoided at all costs. Something you pretend to do and they pretend to pay you for. Or maybe something that was once rewarded. No. The true, precise, unique, and unalterable definition of work is

a force applied over a given distance,

or

$$\text{Work} = Fd. \qquad (2)$$

Well, that was enlightening, wasn't it? Actually it was, you just don't realize it yet. Everyone knows that if you push a large mass (perhaps yourself) over a distance, you feel it, and the larger the mass or the farther you push, the sooner you say, "To hell with this," and get a beer. Equation (2) is the physicist's way of quantifying such frustration. We can make this a little clearer if we rewrite (2) using Newton's second law, $F = ma$. Then the definition of work becomes, inevitably,

$$\text{Work} = \text{mad} \qquad (3)$$

and where the dependence on mass is now explicit.

Now, if you recall from Chapter 1, mass is something not really defined in physics, so work and hence energy are also meaningless. However, although the definitions may be meaningless, they are useful. Suppose a weight is falling through a height h under the pull of gravity. Recall, the acceleration of gravity (9.8 m/s^2) is usually denoted by g. Thus (3) becomes

$$\text{Work} = mgh. \qquad (4)$$

We refer to the quantity mgh or mad as potential energy. Potential energy is often termed the **energy of position**. A Wall Street stockbroker hurling himself from a window has a large potential energy that can flatten a pedestrian—or be harnessed to power a lightbulb. Joule's weights falling from the top of his contraption also have a potential energy, given by (4). It was this potential energy that was used to spin the paddle wheels and heat up the water. In Joule's honor the scientific unit of energy is termed the joule (J). Ah, but we have just said that heat is a form of energy. This means a calorie must also be a unit of energy. True. In fact, what Joule really showed was that it took 4.186 joules to heat up one gram of water one degree, or that one calorie equals 4.186 joules.

Potential energy—the energy of position—is one of the two types of energy distinguished by physicists. The other is **kinetic energy**, the energy of motion. On page 31, the useful formula (10) gave the velocity picked up by an object traveling over a distance d at an acceleration a: $v = \sqrt{2ad}$. This can be rewritten as $v^2/2 = ad$. Substituting $v^2/2$ for ad in (3) above gives immediately:

$$\text{Work} = (\tfrac{1}{2})mv^2. \qquad (5)$$

The quantity $(\tfrac{1}{2})mv^2$ is the kinetic energy of an object traveling at velocity v. And this famous formula tells us that such an object can do an amount of work precisely equal to its kinetic energy. Sometimes (5) is referred to as the "work-energy" theorem. **An object's total energy is equal to the sum of its potential and kinetic energy.**

Now Joule's experiments showed that the potential

DEMO 1

Bernoulli's Derivation of Boyle's Law !

You require Newton's second law and almost all the definitions from Chapter 1.

In Chapter 2 we mentioned that Daniel Bernoulli in 1738 derived Boyle's law on the assumption that pressure was caused by the action of microscopic particles hitting the walls of a container. Indeed, he got more than Boyle's law. He also stumbled upon the ideal gas law, but we'll stick to Boyle's. If you understand this demo, you have passed the first three chapters. If you don't get it, take comfort that it's just a sidebar. This is also a great example of how simple reasoning from basic definitions can lead to profound results.

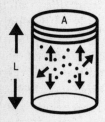

In a cylinder with a piston head, one expects that about ⅙ of the particles are moving up at any time.

Consider a cylindrical container of height L, with a movable piston of area A for a top. Following Bernoulli, we'll assume the cylinder is filled with gas containing N particles, bouncing around. Some particles will be moving up, some down, some east, some west, some north, some south. (And for all we know, some may be going to Daytona for Spring Break.) Since there are six possible directions, let us assume that $N/6$ particles are moving up at any time. As these $N/6$ particles strike the piston, they produce a pressure on it, P. Our mission, should we choose to accept it, is to calculate P. (We'll grudgingly admit that the follow-

DEMO 1

(continued)

ing would never have made it as an episode on *Mission: Impossible.*)

From Esoteric Terms (page 44), pressure is defined as the force per unit area, so let us calculate the force, F, on the piston. The strict definition of force is the change of momentum per unit time (see Talmud, p. 34). If each particle has a mass m, and a velocity v, its momentum is mv. As it strikes the piston, the particle reverses direction and its velocity changes from $+v$ to $-v$, a change of $2v$. The particle's momentum therefore changes by $2mv$. Momentum is usually denoted by the letter p and we denote the change in momentum by $\Delta p = 2mv$.

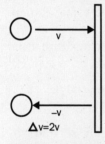

When a particle strikes a wall and reverses direction, its change in velocity is 2v.

We now have the change of momentum of one particle. If $N/6$ particles strike the piston, the total change of momentum is then $N/6 \times 2mv = (N/3)mv$. But force is the **rate of change** of momentum, that is, the momentum transferred to the piston each second. So we now need to figure that out.

Suppose after a time t seconds, all $N/6$ particles moving up strike the piston and transfer the total momentum $(N/3)mv$. That means that the momentum transferred each second is simply $(N/3)mv/t$. To be concrete, say, every $t = 5$ seconds, all the particles moving up strike the piston. The momentum transferred each second is merely $(N/3)mv/5$.

DEMO 1

(continued)

To calculate the rate of change of momentum, all we have to do then is figure out t, the number of seconds it takes for all $N/6$ particles to strike the cylinder. But that's easy. The length of the cylinder is L. As always distance = velocity × time. Thus if a particle is traveling at velocity v, then the number of seconds it takes to cross the cylinder is just $t = L/v$. After this time, all $N/6$ particles moving up should hit the piston and transfer the total momentum $(N/3)mv$.

Usually the change of momentum per unit time is written as $(\Delta p)/(\Delta t)$, so we have:

$$(\Delta p)/(\Delta t) = (N/3)mv/t = (N/3)mv/(L/v)$$
$$= (N/3)mv^2/L. \qquad (1)$$

But the change of momentum each second, by definition, is the force on the piston, just what we want:

$$F = (N/3L)mv^2. \qquad (2)$$

Pressure, by definition, is force per unit area. The area of the piston is A, so $P = F/A$. So divide both sides of (2) by A to get

$$P = (N/3LA)mv^2. \qquad (3)$$

But what is LA? This is just the cylinder's height times the area of its base, in other words, its volume: $V = LA$. So,

$$P = (N/3V)mv^2. \qquad (4)$$

DEMO 1

(continued)

or, finally

$$PV = (N/3)mv^2. \qquad (5)$$

Now, N is the total number of particles in the can, and it doesn't change. Neither does m change. Thus, if we assume v is fixed, we get

$$PV = \text{constant} \qquad (6)$$

which is nothing other than Boyle's law! This all came from the mere assumption that there are molecules in the can, the application of Newton's second law, and the definition of pressure.

---

energy of the falling weights could be transformed into the kinetic energy of the revolving paddle wheels and from there into heat. And that in the transformation process none was lost. This is an example of one of the most important of all natural laws, the

> **Conservation of Energy:** Energy is neither created nor destroyed, merely changed in form. Conservation of energy is so important that it is termed the first law of thermodynamics.

Sometimes it is also known as

> **The First Law of Thermogoddamnics:** You can't win.

Why such optimism? The message of conservation of energy is that you can't get something for nothing. The

Esoteric Terms
(es-ə-'ter-ik tərms)

- *Joule*—The physicist's unit of energy. Energy = *Fd*, and so a one-newton force applied over a distance of one meter is defined to be one joule. Also known as a newton-meter.

- *Calorie*—An infidel unit of energy. One calorie equals 4.186 joules.

- *Specific heat*—The amount of heat energy needed to raise one gram of a substance by one degree Celsius.

- *Potential energy*—The energy of position. An object at height *h* above the ground has potential energy of *mgh*, where *m* is its mass and *g* is the acceleration of gravity. With that certain gram of substance mentioned above, you'd have enough energy to play the position of center for the Celtics.

- *Kinetic energy*—The energy of motion, equal to $(\frac{1}{2})mv^2$.

- *Fermi energy*—We haven't mentioned it, but it's always nice to know there is more to learn.

- *Entropy*—As used by the public it has no meaning. As used in thermodynamics, it is a quantity, similar to energy, that measures the waste heat lost in irreversible processes. In fact, in many situations, it is just the heat transferred divided by the temperature: $\Delta Q/T$.

suicidal stockbroker had a potential energy equal to *mgh*, where *h* was the height of the fatal window above the ground. As he fell, his height from the ground decreased, so his potential energy decreased, but he fell faster and faster, so his kinetic energy increased. His total energy was always the same. When you turn on a lightbulb, the amount of energy in the light plus the heat given off by the bulb equals the amount of electrical energy coming into the socket, which in turn equals the amount of energy given off by the coal at the electrical plant. Conservation of energy tells us that it is impossible to make an engine that produces more energy than it uses up. Misguided souls have been trying to build such **perpetual motion** machines since time immemorial. It was impossible then, it's impossible now, and it always will be impossible. Conservation of energy cannot be violated.

Interestingly enough, for such a famous law, no single name is attached to it. In the mid-1840s, about half a dozen researchers, including Joule, came to the conclusion that energy was conserved and announced the law independently of each other. Thus you trivia fanatics and listmakers are deprived of a historic moment and are forced to actually learn something.

MORE FRICTION, MORE FUN

Conservation of energy is not the whole story. At the turn of the nineteenth century there was great interest in steam engines. In a steam engine—as in a gasoline engine—you may have a piston that returns to the same position after each cycle, ready to start over again. Noticing this, scientists of the time asked a really stupid question: why do you need a continuous supply of coal to run the thing? That is, if energy is conserved, can't you burn coal, produce heat, boil water, make steam, use the steam to run a locomotive, then convert the energy produced by the locomotive into heat again and keep running the engine? In doing so you would create a perpetual motion machine, simply by recycling the output of an engine to its input.

Perpetual motion machines that produce more energy than they use are called perpetual motion machines of the first kind; they are ruled out by conservation of energy. But here, we are obeying conservation of energy. Perpetual motion machines that do not produce more energy than they use are termed perpetual motion machines of the second kind.

Centuries of failed attempts to build one suggested that perpetual motion machines of the second kind were impossible, but no one knew why. In 1803, one of Napoleon's generals, the engineer and mathematician Lazare Carnot, concluded that the most perfect engines—the most nearly "perpetual"—were those that minimized the friction between parts. Friction generates heat that is wasted and doesn't go into producing work, so another way of stating Carnot's observations is that the best engines are those producing the smallest amount of waste heat. But all real engines are subject to some friction and generate some waste heat, so Carnot made the bold inference that perpetual motion machines of the second kind were impossible.

In 1824, Lazare's son, Sadi, tried to prove the assertion. He devised an idealized steam engine. You couldn't actually build the thing, but it was the most perfect engine one could imagine. As the piston moved from one position to another, it lost no heat. Nevertheless Sadi claimed to prove that you could never get as much work out of this perfect engine as the amount of heat you extracted from burning coal.

It turned out that Sadi's proof was flawed, but the conclusion was so obviously true that William Thomson, better known as Lord Kelvin, postulated it as a law of nature. He said simply:

> Perpetual motion machines of the second kind are impossible.

This was the first statement of the infamous **second law of thermodynamics**.

Now, steam engines run by extracting heat from steam; in the process the steam cools and condenses into water. If the cool water in the condenser could spontaneously put back heat into the hot boiler, more steam could be produced and the engine could run forever. This never happens. Thus the impossibility of perpetual motion machines of the second kind is equivalent to a version of the second law stated by Rudolf Clausius:

> Heat cannot spontaneously flow from a colder body to a hotter body.

It was also Clausius who formulated the second law in its most notorious version. That heat flows spontaneously from a hot body to a cold body but not vice versa is an example of an **irreversible** process. When coal is burned it produces ashes; we never see the ash reconstitute itself into coal. When a steam engine operates it loses heat. That heat is gone forever and cannot be recycled. Indeed, in any irreversible process, some useful energy is lost in the form of waste heat. In 1865 Clausius introduced a quantity to measure how much waste heat is produced in irreversible transformations. He termed that quantity **entropy** and rewrote

> **The Second Law of Thermodynamics:** The entropy of an isolated system never decreases.

Here, isolated refers to a system that exchanges neither energy nor matter with the outside world.

The "never decreases" caveat in the second law refers to reversible processes, such as a perfect engine. In such situations no heat is lost and the change in entropy is zero. But that is an idealization. In any realistic, irreversible process, you always produce waste heat, and this amount is measured as an increase in entropy. In other words, a perpetual motion machine of the second kind would have to be a perfectly reversible engine. They don't exist. It is hardly surprising that the second law often goes by the name of

> **The Second Law of Thermogoddamnics:** You can't break even.

One can show that, in a real sense, an increase of entropy corresponds to an increase of disorder in a system. That is, an egg in the hand is an ordered system; an egg smashed on the floor is a disordered system. One can legitimately say that the entropy of a smashed egg is higher than the entropy of the whole egg. For this reason you may well have heard the word "entropy" used as a synonym for "chaos" or "confusion." Indeed, the second law of thermodynamics is certainly the most popular concept from physics to hit the public. However, in all such transitions, entropy increases. If you really want to see a physicist get angry, start using the second law to explain economics, politics, love, the state of the world, etc., as illustrated in the Cocktail Party Conversation below.

UNMAKING OMELETTES

The fact that we never see an omelette reconstitute itself into a fresh egg is perhaps the most puzzling and pro-

COCKTAIL PARTY CONVERSATION

A physicist is wearing a "Maxwell's Equations" T-shirt. You, wearing a "Nuke the Unborn Gay Whales for Jesus" T-shirt, approach.

YOU: Tell me, aren't physicists being a bit pig-headed in refusing to recognize the implications of the second law?

PHYSICIST: Well, we discovered the second law. But what implications did you have in mind?

YOU: I mean, isn't it true that every beer drunk now means one less bottle in the future?

PHYSICIST: If you mean the beer supply is running out, I don't follow.

YOU: Entropy is increasing; you've decreased the available energy. Entropy always gets you. It—it's like karma—what goes around, comes around.

PHYSICIST: But . . . but the second law only applies to isolated systems—systems that don't get any energy—like from the Sun.

YOU: You know, the world sociopolitical situation is showing alarming entropy increase—

PHYSICIST: But . . . but how do you measure it? . . . Political temperature . . . yes . . . I see. . . .

YOU: Entropy. There's no way around it. Society must minimize its energy flow. . . .

PHYSICIST: Vocabulary! Energy or entropy?

YOU: You know, the only way to beat karma is love. Love is anti-entropic, becoming as opposed to being. Nirvana is the state of lowest entropy. . . .

PHYSICIST: Mayday! Mayday! Am being–becoming overwhelmed by cardiac entropy! Beam me up, Scotty!

COCKTAIL PARTY CONVERSATION

(continued)

YOU: There is no way out. . . . To lower entropy we must all become Buddhists.

PHYSICIST: Stardate 2382.07. We have reached Dhyana Four, the Fourth Meditative State, beyond the limits of known thought. Must respond to an urgent call from Nirvana. The beer supply has run out. Transporter!

(The physicist vanishes in a puff of smoke.)

found fact in nature. Why? Because the increase of entropy defines a direction in time. "The future" is the direction in which entropy increases. This would not be so mysterious if, according to Newtonian mechanics, it did not seem impossible.

Recall from Chapter 1, we said that given the initial position and velocities of all objects in a system, Newton's second law could predict the entire future behavior of the system. This works in reverse as well. Merely by letting t in the equations go to $-t$, one can determine exactly where the system was at any time in the *past*. As an example, suppose you shoot a film of the planets orbiting the Sun. If you run the film in reverse (the same as letting t go to $-t$), the planets move backward and you might be tempted to say time is running backward as well. But how do you know I didn't shoot the film backward to begin with? In Newtonian mechanics the equations work equally well running forward or backward. There is no distinction between past and future. Newtonian mechanics is **time reversible**.

And this brings us to the central paradox. If the atoms in an egg or a lump of coal obey Newtonian mechanics, then they do not distinguish time forward from time backward. Yet, we do not see omelettes unscramble or coal unburn. Nature is **irreversible**. How can we reconcile the second law with Newtonian mechanics? The great physicist Ludwig Boltzmann thought he could derive the second law from Newtonian mechanics. He was mistaken (it is mathematically impossible to derive a time irreversible law from time reversible laws) and the attacks on his proof helped drive him to suicide. That two great pillars of physics, Newtonian mechanics and thermodynamics, seem to stand on irreconcilable foundations is perhaps the most profound problem in physics and the subject of much debate even today.

SUMMARY

⏱ Key Words: Heat, energy, entropy.

⏱ Key Definitions: Potential energy = mgh; kinetic energy = $(\frac{1}{2}) mv^2$; entropy measures the amount of energy wasted in any real process.

⏱ Key 'cept: The laws of thermodynamics are at the bottom of the environmental crisis.

CHAPTER 4

ELECTRO-MAGNETISM:

A CURRENT AFFAIR

YOU MUST REMEMBER THIS

The forces of electricity and magnetism appeared related but distinct until the mid-nineteenth century when Maxwell showed they were indeed two aspects of the same phenomenon. In doing so, he created the first unified field theory. Also, the work of the nerds who invented the science of electrodynamics underlies much of modern technological civilization.

ALL CHARGED UP
AND NOWHERE TO GO

You have probably never stopped to consider the connection between magnets and laxatives, but there is at least one: The word "magnet" comes from Magnesia, as in "Milk of Magnesia," Magnesia being a region of ancient Greece now located in Turkey. Magnesia was rich in lodestone, a naturally magnetic mineral, and presumably in the element magnesium. The great thing about language is that every word opens a tiny window onto the past. "Electricity" is also from the Greek, *electrum*, meaning "amber."

Amber? Well, it appears that the ancient Greeks— some say Thales himself—knew that just as lodestone attracts bits of iron, amber, when rubbed by fur, attracts bits of straw. The similarities between electricity· and magnetism undoubtedly attracted the attention of natural philosophers for centuries. The investigation of their relationship resulted in the science of **electrodynamics,** which is brought into play every time you turn on a lightbulb, a television, or an electric toothbrush. It also produced Einstein's theory of relativity. But before all that, the most striking things about electricity and magnetism were not their similarities but their differences.

You can discover the main difference yourself by some simple experiments. The first was performed by, among others, Benjamin Franklin. Besides guiding Declarations of Independence, charming Frenchwomen, and inventing stoves, Franklin was in fact a first-rate experimenter. In this case, Franklin rubbed a glass rod suspended from a thread with a silk cloth. Stroking a second glass rod with silk, he brought it near the first and found that they repelled each other. But if he rubbed a hard rubber rod with fur, he found that it *attracted* the suspended glass rod. By such experiments Franklin concluded that electric "fluid" (yes, phlogiston and caloric were not the end of the story) contained two types of electric charge, which he termed **positive** and **negative.** At the same time he deduced the familiar love–hate relationship between charges:

> **Electrical Love–Hate:** Like charges repel; unlike charges attract.

Now, take two bar magnets and hang one from a thread. As you probably know, a bar magnet has a north pole and a south pole. If you bring the second magnet near the first you will discover

> **Magnetic Love–Hate:** Like poles repel; unlike poles attract.

Indeed, by hanging a magnet from a thread you have followed the ancient Chinese in constructing a compass. Because of the love–hate relationship, it is actually the south pole of a compass that points to the north magnetic pole of the Earth and vice versa.

Yet it would be a mistake to conclude electricity and magnetism are totally analogous. By rubbing a glass rod with silk we say you have **charged** the rod. A dangerous word. It does not mean you have created any charge; as Franklin realized, it only means you have separated the positive and negative charges. If the rod becomes positively charged, the silk becomes negatively charged. In fact no experiment has ever been done or ever will be done that actually creates charge. This truth is expressed in our next great Commandment, proposed by William Watson in 1746 and Franklin in 1747:

> **Conservation of Charge:** Electric charge is neither created nor destroyed. The total amount of charge in the universe is constant.

On the other hand, can you separate north and south magnetic poles? If you break a magnet in half, what do you get? A north pole and a south pole? A refund? No, of course not. You get two new magnets, each with a north and a south pole. It is a profound fact of nature that it is possible to isolate the two types of electric charges but that it is impossible to isolate magnetic poles. In fact, let us call this

> **The Monopole Law:** There are no isolated magnetic poles in nature.

WHO'S WHO

Benjamin Franklin (1706–1790)
Early experiments with electricity. Named the two types of electrical charges, but got them backward. Flew a kite. Helped found United States. Judging from his exploits as an ambassador to France, bald was sexy.

Charles Coulomb (1736–1806)
Originator of the term "charge it." Measured the attraction or repulsion between two electric charges, and found they obeyed an inverse-square law. Unit of measure named after him: the "coulomb," a unit of electric charge.

Joseph Henry (1797–1878)
Did all the calculations, got no credit; did get unitized as "henry," the unit of inductance. Was first director of Smithsonian. Second big American scientist, behind Ben Franklin.

André-Marie Ampère (1775–1836)
Overcome operatic family life, and discovered

WHO'S
H
O
☛

(continued)

how electric current produces magnetic field. Unit named after him: the "amp," the unit of electrical current.

Michael Faraday (1791–1867)
Discovered Henry's law of inductance, which shows how changing a magnetic field produces electric current; got unitized as "farad," the unit of capacitance.

Thomas Young (1773–1829)
Demonstrated light is a wave; did not get unitized, but deciphered demotic script on the Rosetta stone.

Hippolyte Louis Fizeau (1819–1896)
Provided first decent measurement of speed of light. Had no unit named for him, which is sad, given the poetic aptness of the unit that never was: a "fizz" of light.

James Clerk Maxwell (1831–1879)
Married electricity and magnetism into electromagnetism. Considered the greatest physicist of the nineteenth century, but for some reason, he too missed the units' hall of fame.

Heinrich Hertz (1857–1894)
Discovered Henry's radio waves; got unitized as "hertz," the unit of electromagnetic wave frequency. No obvious connection to car rentals.

Kayser and **Talbot**
No one's ever heard of them, but they've got units.

COULOMB LAYS THE LAW

Because natural philosophers are disciples of the **quantification principle** (you don't know anything until you've assigned a number to it) it was not enough to know that charges attract or repel. The question was, what is the strength of the attraction or repulsion? Experiments in 1772 by the same Cavendish who investigated the composition of water and by Charles Coulomb in 1785 resulted in what is not called Cavendish's law (since he perished, not published), but Coulomb's law. The force of attraction or repulsion between two electric charges is proportional to the product of their charges and inversely proportional to the square of the distance between them:

$$F_{ELEC} = k_c q_1 q_2 / r^2. \qquad (1)$$

Here, q_1 and q_2 represent the two charges and r the distance between them; k_c is a constant. If you compare this with the formula on page 36 for the gravitational attraction between two masses, you will see they are virtually identical. They are both inverse-square laws; if you double the distance between the two charges, the force decreases by a factor of 4, etc. The only real difference between the two formulas is that here we have charges (q's) instead of masses (m's). As we know, mass is measured in kilograms, but charge is measured in a thing called the **coulomb**. (It has been said that the highest honor in physics is to become a unit of measurement; for the definition of coulomb, see Esoteric Terms.) The fact that we have here q's instead of m's is essential: m's are always positive, so the gravitational force is always "positive" (attractive), whereas q's can be positive or negative. That means the electric force can be attractive or repulsive. It also should go without saying that there is no electric force between uncharged objects, since in that case the q's are zero.

Then we have this k_c instead of G. Both are just numbers that must be measured in the lab. G, the gravitational constant, is the number that fixes the size of the force of gravity. The number k_c does the same thing for the electric force. (Its numerical value is 9×10^9 in the units we have been using.) The important thing is that the electric force is enormously stronger than the gravitational force. In fact the electric force between an electron and a proton (the charged particles in the atom) is about 10^{39} times stronger than the gravitational force!

The huge size of the electric force is *prima facie* evidence that the total electric charge in the universe is zero. For example, in your body there are about 10^{29} protons and electrons. If one out of every 10,000 of your body's electrons somehow migrated to the center of the Earth, leaving you with a small excess of protons, you would be smashed to the ground by a coulomb force one million times the force of gravity! Here is good reason to believe you are electrically uncharged, or **neutral.** *And an even better reason to stay that way.*

This exercise shows why Coulomb's law is enormously important in physics—and daily life. It is the coulomb force that holds electrons in orbits around the nuclei of atoms. It is the coulomb force that holds atoms together in molecules and it is the coulomb force that binds molecules together into liquids or solids. There is, however, one species of bond in which even the coulomb force cannot hold opposing elements together: the celebrity marriage.

GOING WITH THE CURRENT

Two immobile charges are termed, not surprisingly, **static,** and apart from gravity the only force they feel between them is the coulomb force. Frankly, static electric charges have limited audience appeal. But things really get exciting when the charges start moving to form an electric **current**. It was by observing electrical currents that the connection between electricity and magnetism was irrevocably established. In 1820 Hans Christian Oersted (1777–1851) of Denmark found that a wire carrying an electric current deflected a nearby compass needle. Today an eight-year-old can do the experiment in about five minutes with an ordinary battery, a wire, and a compass. What's the big deal? In those days, the demonstration probably would have won a Nobel prize, had they existed. One can see the tabloids now:

MOVING ELECTRIC FLUID PRODUCES MAGNETIC FIELD!
WHAT DOES IT ALL MEAN?
PHILOSOPHERS SUSPECT SPACE ALIENS.

Probably not. As to what it all means, a large step was taken by the fellow you remember every time you replace a blown fuse: André-Marie Ampère. His immortalization as the unit of electric current, the **amp,** disguises a different life. Ampère's father was guillotined in the Reign of Terror, his beloved first wife died in 1803, his second marriage ended in divorce, his daughter married an insane, alcoholic officer in Napoleon's army, and his son wasted twenty years as a courtier of the famous Mme. Récamier.[1] Nevertheless, Ampère managed to discover the law that related electric currents to magnetic fields. That is, he quantified Oersted's observation.

Because we have foresworn calculus in this book, it is difficult even to write down Ampère's law properly. But this law tells you precisely how much magnetic field is produced by a given electric current. For Oersted's demonstration, in which a current traveled through a single wire, Ampère's law says that the magnetic field at a given

---

[1]Pearce Williams, "André-Marie Ampère," *Scientific American* (Jan. 1989).

point in space is proportional to the current and inversely proportional to the distance from the wire:

$$B = k_B \times I/r \qquad (2)$$

where we have followed the physicist's obvious mnemonic device of using B for the strength of the magnetic field and I for the current; k_B is yet another proportionality constant, a number that you have to measure in the lab. The moral of Ampère's law is: **Electric currents produce magnetic fields.**

Joseph Henry— Discovered magnetic inductance before Faraday, radio waves before Hertz and invented the telegraph before Morse. Rarely credited, but today we call the unit of inductance a "henry."

Well, if that is true, you might wonder if the reverse also holds: can magnetic fields somehow produce electric currents? The answer is emphatically yes. The method by which magnetic fields produce electric currents is termed Faraday's law of induction after Michael Faraday, and was discovered by Joseph Henry in 1829 (see box on page 97). You can duplicate rather easily what Henry and Faraday did with some wire, a bar magnet, and a multimeter (set on the most sensitive current scale). Wrap the wire into a coil around a cardboard tube and attach the ends to the meter. Now thrust the magnet through the coil. You will see the needle of the multimeter momentarily register a current, then swing back to zero.

By moving the magnet through the coil, you have **induced** a current in the wire. But the current stops when the magnet stops moving. This is the key observation of Faraday's law: **A changing magnetic field produces an electric current.**

Just to get an idea of what it looks like, we can write Faraday's law for a coil in a somewhat simplified form as:

$$I = \text{constant} \times \frac{(\Delta B)}{(\Delta t)} \qquad (3)$$

where $(\Delta B)/(\Delta t)$ means the rate of change of the magnetic field. The law of induction is the basis for the transformer and is at the bottom of much of the power

JOSEPH HENRY

Joseph Henry grew up in Albany, New York, where he attended school at the Albany Academy and, among other things, spent several years as an apprentice actor. In 1826 he became Professor of Natural Philosophy and Mathematics at the academy, and during summer vacations carried out his experiments on electricity and magnetism. In 1829 he made his great discovery that a changing magnetic field can induce an electric current in a wire. Unfortunately, as a provincial American, he was unaware that others in Europe were working along the same lines and delayed publication for several years, during which time Michael Faraday made the same discovery. In 1837 the two men met. Faraday (often called "the greatest experimentalist of all time"), already credited with the discovery of inductance, needed Henry to explain it to him.

Henry also invented the telegraph key about five years before Samuel Morse, invented the transformer, and discovered radio waves in much the same way as did Heinrich Hertz—but a half century earlier. He also recognized that radio waves travel at the speed of light and he was the first to measure the temperature of sunspots. In 1832, Henry became professor at New Jersey College, now Princeton University, and later the first director of the newly founded Smithsonian Institution. Probably the greatest American scientist after Franklin, Henry is immortalized by the "henry," the electromagnetic unit of inductance, but he is still infrequently mentioned in textbooks.

industry; you use it each time you charge your Interplak toothbrush. Indeed, we have it on unimpeachable authority that when Faraday was asked by a woman of what use was induction he replied, "Of what use is a newborn babe?" Other unimpeachable authorities say that it was a government official who asked and Faraday replied, "I don't know, but someday you will tax it."

FORCE PLAYS AND FIELDERS CHOICES

Now, your author has been glibly using the term "magnetic field" without defining it because the expression has entered the English language, along with "gravitational field," "electric field," "field of influence," and many other types of less-definable force fields available to consumers in the New Age. But though actors bandy about such jargon on every episode of *Star Trek*, you can be certain neither they nor the writers have the faintest idea of what the words mean. The idea of a field came about because of conceptual problems inherent in expressions like Newton's law of gravity and Coulomb's law, which can describe the forces between two objects separated by great distances.

Natural philosophers, including Newton, had a hard time accepting that a force could be transmitted across empty space. Eventually Faraday introduced a new abstraction to get around this problem. He visualized a **field** that filled all space with **lines of force** that told an object in what direction to move and with what acceleration. The field was meant to be the **medium that conducts the message.** At that time, it was a field of dreams. The conceptual hurdle has vanished over the centuries because if you know the strength of the field at any point, you know the force acting on an object, and vice versa. It really doesn't matter whether you speak of

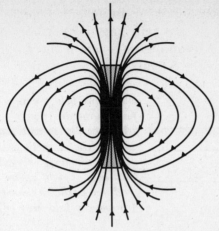

The magnetic field lines around a bar magnet give rise to the magnetic force, which acts on particles.

forces or fields. For example, if at a certain location gravity is pulling an object of mass m with force F, then the field strength at that point is defined as $g = F/m$. In other words, field strength is just the acceleration. By analogy, for electric forces acting on charges, the electric field E is defined as $E = F/q$. That means in Coulomb's law (1), the field strength on the charge q_2 is just given by $E = k_c q_1/r^2$. (For more on the relation between forces and fields, see the feature on page 101).

For magnetic fields, the story is a little trickier. It turns out that magnetic fields deflect charged particles *at right angles* to both their direction of motion and the direction of the B-field itself. For a particle moving at right angles to the B-field, the force acting on it is $F = \pm qvB$, where v is the velocity, relative to the direction of the magnetic field. (Whether the + or − is chosen depends on the direction of the particle's motion relative to the magnetic field.) With this definition of force, the magnetic field strength is $B = F/qv$. Indeed, the total force acting on a charged particle *moving at right angles to a magnetic field* is given by the important formula:

$$F = qE \pm qvB. \qquad (4)$$

We will have occasion to use this formula in Chapter 6. Note that F is zero if q is zero (neutral particles not deflected). Also, the second term is zero if the particle is moving parallel to the B-field.

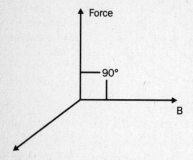

An electric field produces a force in the direction of the field lines. Not so with magnetism. The magnetic force (F) on a moving charged particle is at right angles to both the particle's velocity (v) and the field lines (B).

Now, we have been talking a lot about electric currents. Currents are just charges in motion. They are put in motion by forces—or fields. You can easily believe that a large electric field in a wire will produce a large current. In that case, (2) and (3) are really telling us something about the relationship between electric and magnetic **fields.** Indeed, we might rewrite (2) and (3) somewhat loosely as:

$$B = k_B \times E/r \qquad (2)$$

and

$$E = \text{constant} \times (\Delta B)/(\Delta t). \qquad (3)$$

Where Δ means "change in" you can regard Ampère's law (2) as telling you how much B-field is generated by an E-field, and you can regard Faraday's law (3) as telling you how much E-field is generated by a changing B-field.

DEMO 1

More on Forces and Fields !!

Make sure you know definitions of force and field strength from the text. You also need the definition of potential energy from Chapter 3.

Phycists designate the gravitational field strength to be defined as $g = F/m$; it is just the acceleration, g. By analogy the electric field strength is defined as $E = F/q$. Notice the analogy is not exact. We know by now $F = ma$, *always*. That means the electric field is $E = F/q = ma/q$. The field strength is then *not* the acceleration, as it is for the gravitational field, but the acceleration multiplied by the ratio (m/q).

Also, we defined potential energy (Chapter 3, Equations 2 and 4) as a force applied over a given distance, $PE = Fd$. For gravity the force on a mass is, from above, $F = mg$. For an electric field the force on a charge is $F = qE$. We can then write for the potential energy of the charge

$$PE = Fd = qEd. \qquad (1)$$

The combination Ed is given a special name: volts; $Ed = V$. Now the idea of an electric field can be made a little less abstract. Connect a one-volt battery across two electrodes separated by a distance d of one meter. Since $Ed = V$, and $E = V/d$ and the battery has set up an electric field of one volt per meter. Indeed, we usually think of electric fields as so many volts per meter, rather than so many newtons per coulomb (given by the definition $E = F/q$).

DEMO 1

(continued)

In terms of volts, the potential energy (1) can be rewritten as

$$PE = qV. \quad (2)$$

This is why one often hears volts referred to as "electrical potential." Equation (2) also gives rise to a new unit of energy. We say that if an object's charge q is the charge on one electron (1.6×10^{-19} coulombs) and V is one volt, then the object's potential energy is *one electron volt*. Because of the way a coulomb is defined, it turns out that one electron volt is also 1.6×10^{-19} joules—a very small number! The electron volt (eV) is a convenient unit of energy in atomic physics because many processes involve energies of about 1 eV. Usually the energy produced by particle accelerators (Chapter 8) is expressed in electron volts.

DOING THE WAVE

We must now take a pseudo-digression, a digression that appears to be a digression but is not a digression. This pseudo-digression concerns the nature of light, a topic that occupied natural philosophers for centuries. Newton himself, under the influence of Boyle, believed light was composed of minute particles like billiard balls. His rival, Christian Huygens, argued more convincingly that light was composed of waves.

Waves are so important that neither physics nor society could operate without them. The most familiar type of wave is probably a water wave. If you drop a pebble into a pond, waves spread outward. If a cork is sitting on the pond's surface, it will bob up and down as the ripples pass, but it only moves up and down—not sideways. This shows that the water is not moving past—the wave is. The same phenomenon shows up more vividly on the *New York Newsday* update on Times Square, where the latest news ELVIS SIGHTED IN LENIN'S TOMB moves around the perimeter of the building, written in so-called "traveling lights." Actually, the lights aren't traveling at all; they are merely blinking on and off in succession; a wave—ELVIS—is traveling, not the lights. Here is a case where the medium is definitely not the message.

Waves have a velocity, usually denoted by c, and they can transmit energy. Indeed, every time you turn on a radio, radio waves are depositing energy in the receiver. In describing waves we often speak unimaginatively of their **amplitude** (their height), their **wavelength** (the distance between two adjacent crests), and their **frequency** (the number of crests passing a given point every second) (see figure 2). From these definitions, it is easy to show for any wave the extremely important relationship

$$c = \lambda\nu, \qquad (5)$$

where λ (lambda) represents the wavelength and ν (nu) the frequency. Notice that if the velocity of the wave c is fixed, as it usually is, then as the wavelength of a wave goes up, frequency must go down and vice versa. So you can talk about the wavelength of a radio station or its frequency, it doesn't matter which. Equation (5) lets you convert from one to the other.

One of the most important features of waves is their ability to undergo **interference.** That is, two waves traveling in opposite directions will pass through each other

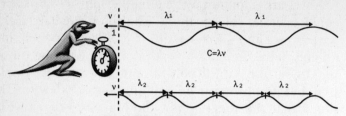

Basic Wave Diagram—

A creature observes two waves passing by its eye at a velocity, c. The distance between any two crests is termed the **wavelength,** *denoted by* λ. *Also, the number of wavelengths passing the creature's eye each second is termed the* **frequency** *denoted by* ν. *If in (a), the creature sees two wavelengths passing its eye each second, it says the frequency is 2 cycles per second. Thus it takes only ½ second for the wave to travel* **one** *wavelength. The time to cross one wavelength is termed the* **period,** *usually denoted T. In (b) the wavelength is half as long as in (a). Therefore four crests pass the creature's eye each second, the frequency is 4 cycles per second and the period is ¼ second. Note that the shorter the wavelength, the higher the frequency. Note also that the period is the reciprocal of the frequency: T = 1/ν.*

Now, the waves are traveling at velocity c. As always, distance = velocity × time. The time to cross a wavelength 1 is by definition the period, T, so $\lambda = c \times T$. *But* $T = 1/\nu$, *so* $\lambda = c/\nu$, *or* $c = \lambda\nu$. *This is the fundamental relationship between frequency, wavelength, and velocity.*

and come out unscathed. But while they are passing through, they are interfering: At each point the total amplitude of the disturbance will equal the **sum** of the amplitudes of the two waves. For example, suppose the two waves are identical. Then, in places where two crests coincide, the total amplitude will be **twice** the amplitude of each wave. In places where a crest coincides with a trough, the waves will "cancel out" and the total amplitude will be **zero**. A standard high-school demonstration of interference is carried out in a thing called a ripple tank.

Two rocks dropped in a lake produce ripples that pass through each other but at the same time add their amplitudes to produce a new wave. The familiar process is termed interference.

For our story, the important thing is that in 1800 Thomas Young, an English polymath, demonstrated that light undergoes interference. Clearly, it is hard to imagine particles behaving in this way. No one has ever seen

THOMAS YOUNG (1773–1829)

Born near Somerset, England, Thomas Young was reading by the age of 2, had gone through the Bible by the age of 4, and spoke Latin by the age of 6. (Of course this begs the question, could he have mastered every arcade game in the universe by age 8?) When he finished school at age 13, by then knowing French and Italian as well, he dove into Hebrew, Chaldean, Syrian, Samaritan, Arabic, Persian, Turkish, and Ethiopian. At 19, having studied Newton's works and taught himself calculus, he entered the field of medicine.

Evidently Young was not successful as a doctor. But we can grant him this failure because in 1800 he performed one of the most important experiments of the century. By demonstrating that light undergoes interference (see next page), he disproved Newton's claim that light was composed of particles and showed that it acts like a wave.

He also made contributions to optics and the study of the elastic properties of solids. Nor were his earlier linguistic studies wasted. Young was the first to decipher the demotic script on the Rosetta stone.

billiard balls interfere. So by his historic experiment, Young disproved Newton's idea that light was composed of particles and showed that it acts like waves. A half-century later, Hippolyte Louis Fizeau of France made the first accurate measurement of the speed of light and found it to be nearly the value we accept today, namely, $c = 3 \times 10^8$ meters per second. End of pseudo-digression.

TAKING IT TO THE MAXWELL

We now approach the climax of our electrifying tale. You may have noticed something slightly odd in the previous discussion. From Equation (2) an electric field produces a magnetic field, but from Equation (3), a *changing* magnetic field produces an electric field. Wouldn't the situation be more symmetric if a *changing* electric field could also set up a magnetic field?

James Clerk Maxwell (1831–1879) thought so too and so he modified Ampère's law, Equation 2, to look something like

$$B = k_B \times E/r + \text{stuff} \left\{ \frac{\Delta E}{\Delta t} \right\}. \qquad (6)$$

That is, he added a term proportional to the rate of change of the electric field. At the time, there was no experimental evidence to indicate (6) was right, but Maxwell was following his sense of form. Indeed, one of the most important guiding principles of physics is the

> **Principle of Beauty:** The correct theories are the most beautiful ones.
> If you must choose between experiment and beauty, choose beauty.

Maxwell's artistic sense paid off. By changing Ampère's law he was able to rewrite the equations for the electric

DEMO 2

Young's Experiment: The Experiment That Showed Light Is a Wave !

Consider the arrangement in the figure. Light impinges on two slits or pinholes in the screen B. If light is a wave, it will fan out from each slit, causing the interference pattern shown. On a distant wall, C, you will see a series of dark and light spots. The light spots occur where crests from the two slits fall on each other and add. (The waves are "in phase.") The dark spots appear where a crest from one wave falls on a trough from the second and the two cancel out. (The waves are "out of phase.") You have probably seen such a phenomenon in a swimming pool. The fact that such interference takes place shows that light is a wave, not composed of particles.

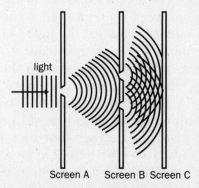

light

Screen A   Screen B Screen C

Thomas Young passed light through one slit
(to collimate it) then through two more slits.
In producing an interference pattern he
demonstrated that light acted as a wave.

DEMO 3

Measuring the Speed of Light !

For this demo you need to know the relationship between distance and velocity from Chapter 1.

In Galileo's dialogue *Two New Sciences,* the protagonists wonder whether it is possible to measure the speed of light. Though we now know the answer is yes, theirs was not a stupid question; the velocity of light is so high that it takes only about a billionth of a second to travel the length of your arm. In the old days, there was simply no reason to think the speed of light was not infinite.

One of the dialogue's personages, who is evidently Galileo himself, describes how he and an assistant attempted to measure the speed of light. Separated by about a mile, Galileo would uncover a lantern. As soon as his assistant saw the light, he would also uncover a lantern. Galileo tried to measure the round-trip time for the light. As far as he could determine, it was instantaneous.

The first measurement that showed the speed of light was actually finite was made in 1675 by Ole Roemer using the moons of Jupiter. When Jupiter is, say, directly overhead, its moons will appear from behind the planet at a certain time each night. But as the year wears on, the moons appear later and later. After half a year the delay is about 16 minutes. Roemer reasoned that this was the time it took light to cross the distance of the Earth's orbit (see figure). If you know the size of the orbit, you can just divide it by 16 minutes and get the velocity of light. There is some confusion over what Roemer actually

DEMO 3

(continued)

did,* but with one set of figures available at the time you get about 1.5×10^8 m/s, half the correct value.

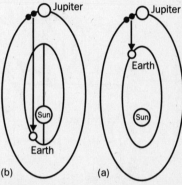

(a) When the Earth and Jupiter are at closest approach, Jupiter's moons appear at a certain time of night. (b) Six months later, when the Earth is at the other side of the Sun, Jupiter's moons appear about 16 minutes later. This delay is due to the time it takes light to cross the Earth's orbit.

The first nonastronomical measurement of the speed of light was made in 1849 by Hippolyte Louis Fizeau. He sent a light beam through the gaps in a rapidly rotating toothed wheel out to a mirror about 8 km away (see figure). Fizeau cranked up the wheel so that, by the time the light returned, the wheel had rotated enough so that a tooth blocked the beam. Since Fizeau knew how fast the wheel was rotating, he could compute the speed of light.

---

*See, for example, Albert van Helden, "Roemer's Speed of Light," *Journal for History of Astronomy* (1983, vol. 14, p. 137).

DEMO 3

(continued)

His answer was very close to the value accepted today: 3×10^8 m/s. Today, students can perform Fizeau's experiment in a freshman lab.

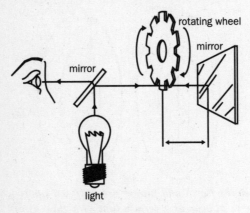

Fizeau's apparatus to measure the velocity of light (see text).

and magnetic fields in a specific way. Mathematically, his new equations yielded an astonishing result: **accelerating electric charges gave off radiation.** This radiation was in the form of **waves** composed of electric and magnetic fields traveling **together.**

But that wasn't all. In equations describing waves, the velocity always appears in a certain place. Maxwell noticed that in that same place he got a combination of the numbers we have been writing as k_B and k_c, more or less $\sqrt{k_c/k_B}$. Remember, these are just numbers that need to be measured in the lab. When Maxwell worked out this combination of k_B and k_c, he found that it equaled exactly 3×10^8 meters per second, the measured value of the speed of light! Maxwell realized that **light was an electromagnetic wave.**

Physicists use the term "light" quite generally. It refers to any electromagnetic wave, and those include visible light, radio, microwaves, X rays, and gamma rays. They are distinguished from each other only by their frequency or—what amounts to the same thing—their wavelength. Taken together, all these waves form the **electromagnetic spectrum.**

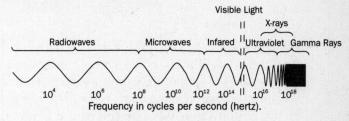

Electromagnetic Spectrum

In showing that electric and magnetic fields were invariably linked in one **electromagnetic field,** Maxwell created the first unified field theory. The goal of combining all the forces or fields in nature into one theory remains with physicists today.

To achieve his results, Maxwell required not only a generalization of Ampère's law, but required Coulomb's law (1), the monopole law (which one can write as an equation), and Faraday's law (3). For this reason, all these expressions are collectively known today as Maxwell's equations of the electromagnetic field.

Maxwell's work, generally considered the crowning achievement of nineteenth-century physics, was theoretical. But the proof of the physics is in the experiment. His prediction of electromagnetic waves was not experimentally confirmed until twenty years later when Heinrich Hertz found that electric sparks in a coil (accelerating charges) at one side of his laboratory were causing sparks in a coil at the other side of his laboratory. As Joseph Henry had done half a century earlier, Hertz had stumbled upon radio waves. The rest, as they say, is history.

Esoteric Terms
(es-ə-'ter-ik tərms)

- *Monopole*—A hypothetical isolated north or south magnetic pole. Thought not to exist in classical physics.

- *Coulomb*—The scientific unit of electric charge. Its formal definition is rather complicated. For our purposes it is perhaps easiest to accept that the electron has a charge of 1.6×10^{-19} coulombs, or conversely that one coulomb is equivalent to the charge on 1.46×10^{19} electrons. That is a large number; a lightning bolt transfers only about 10 coulombs of charge.

- *Coulomb's law*—The law that describes the force between two electric charges; analogous to the law of gravity. Could also have been a TV series about a detective who's forever pressing charges. Here's the math: $F_{elec} = k_c q_1 q_2 / r^2$. The constant k_c has a value equal to 9×10^9 newton-meters^2/(coulomb^2).

- *Ampère's law*—The law that describes how an electric current produces a magnetic field. A TV producer would have changed this to Vampere's law, the shocking story of a battery that runs on blood.

Esoteric Terms
(*continued*)

- *Induction*—Generally refers to the ability of a changing magnetic field to produce an electric current in a wire. The basis of the transformer (and the draft).

- *Faraday's law*—The law that describes how a changing magnetic field produces an electric current. Discovered by Henry.

- *Wavelength*—In the study of electricity (and in waveforms generally), wavelength is defined as the distance between two crests of adjacent waves. At a sporting event, we suppose it's the average span from armpit to hand in· an undulating mass of fans.

- *Amplitude*—The height of a wave, measured either from the center line to a peak or from peak to peak, as you choose.

- *Frequency*—The number of wave crests that pass an observation point every second. Measured in cycles per second, also known as "hertz."

- *Interference*—The summing of the amplitudes or two or more waves as they pass through each other.

- *Field*—An abstract arena on which forces act. The shape and strength of the field at any point determines the direction and size of the force there. Come to think of it, this sounds like a description of Virtual Football.

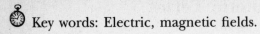

SUMMARY

Key words: Electric, magnetic fields.

Key definitions: The electric force $F = qE$, where E is the field strength. The magnetic force $F = qvB$ for a particle with a charge q moving at velocity v at right angles to a magnetic field B.

Key 'cepts: The forces of electricity and magnetism are part of one **electromagnetic force.** Accelerated electrical charges produce **electromagnetic waves**, otherwise known as light.

CHAPTER 5

SPECIAL RELATIVITY

EINSTEIN SEES THE LIGHT

YOU MUST REMEMBER THIS

Relativity does not mean everything is relative. And the brilliance of Einstein's discoveries is so great that no amount of journalistic overkill has managed to dim it. Einstein and Bach are the only two people who deserve their reputations.

VELOCITY: AN ETHER/OR DILEMMA

Maxwell's discovery that electromagnetic waves travel at the single velocity of light, 3×10^8 meters per second, disturbed him greatly. If you reflect on this fact for a moment, you too will be stricken by doubt and insomnia, for it contradicts everything you know from daily experience.

Reflect: You are playing Mutant Ninja Frisbee with a friend; you toss the Frisbee between you at a certain velocity. To be concrete, we will say 50 kilometers an hour in each direction. Now, suppose an interloper suddenly informs you that you are actually playing Mutant Ninja Frisbee on a train moving at 100 kilometers per hour. You shrug. As far as the game goes, the information is irrelevant: to you the Frisbee always appears to be moving at 50 kilometers an hour. But the interloper's information is relevant to a peeping tom watching from the ground. This spy sees the Frisbee traveling at 150 kilometers per hour in the direction of the train, but at only 50 kilometers per hour against it. From the **frame of reference** of the spy on the ground, one must add or subtract the velocity of the train to the Frisbee.

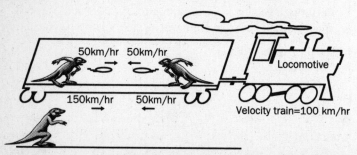

Two creatures play Frisbee on a train. To them the Frisbee appears to fly at 50 km/hr in each direction. However, the train is moving at 100 km/hr with respect to the ground. To a creature on the ground, the Frisbee appears to be moving at 50 km/hr against the train, but 150 km/hr with the train.

The idea that one should add and subtract velocities of moving objects depending on one's frame of reference is due to Galileo and seems quite natural. Of course. But now you see you have been led down the primrose path into paradox and contradiction. Maxwell determined that the velocity of light was 3×10^8 meters per second. Period. Full stop. No mention of any frame of reference. Sound waves travel at 330 meters per second with respect to still air. Water waves travel at a few meters per second with respect to still water. Amber waves of grain travel too, which puts "America the Beautiful" in a whole new light.

But Maxwell's results were different. He didn't find that the velocity was 3×10^8 meters per second relative to the Earth (which is moving around the Sun at roughly 30 kilometers per second). Nor to the Sun itself. As far as his equations were concerned, the speed of light was 3×10^8 meters per second *always*.

This was so hard to believe that Maxwell thought his equations were either incorrect or applied only to one particular reference frame, the reference frame of the "luminiferous ether." The ether was not something you got at a dentist's office. It was yet another piece of unclaimed baggage dropped by Artistotle in the way stations of history. In our Prologue, Anaximander introduced the concept of *apeiron*, "the boundless," from which all other substances emerged. Aristotle modified the idea to suit his own designs. His design, recall, included four elements from which all else emerged. But the heavens were perfect, and therefore demanded their own element. Aristotle called it the **quintessence**, the fifth essence. The ether, as it became known, filled the heavens, completely, immutably.

Natural philosophers took over the idea. In this case the act was not unreasonable. Ocean waves do not roll without water. Sound waves cannot be transmitted through a vacuum; von Guericke and Boyle showed they require air. Why should light be different? The luminiferous ether became the medium that transmitted electromagnetic waves. As such, **the ether defined an absolute standard of rest.**

Here we must pause. Newton's first law states, **"A body moving at constant velocity continues to move at a constant velocity unless acted upon by an outside force."** But we have cleverly ignored the question, "Constant velocity with respect to what?" Newton dismissed the question. And for good reason. Think about it: when you were caught playing Frisbee on the train traveling at 150 kilometers per hour (it had perfectly smooth rails and the blinds drawn) you did not realize you were moving at all. Therefore, **so far as Newton's laws go, there is no way to distinguish a reference frame moving at constant velocity from "absolute rest."**

Unless there was an absolute standard of rest. Unless there was a luminiferous ether. Then a creature sleeping on a train would nevertheless be moving at 150 kilometers per hour with respect to the ether. This velocity would be its absolute velocity, and such motion should be detectable. Relative to this **absolute reference frame,** Maxwell's equations would hold as he derived them, and the velocity of light would be the velocity he computed, 3×10^8 meters per second.

But to the dismay of scientists all experiments designed to detect the ether failed. The most famous of these was carried out by Albert Michelson and Edward Morley, who in 1887 attempted to measure the velocity of the Earth with respect to the ether. The idea was this: The Earth is in orbit around the Sun and is therefore traveling in different directions at different times of the year. Michelson and Morley could thus shine a light beam along one direction with respect to their labora-

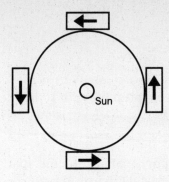

Suppose a lab on the Earth is moving at velocity v with respect to some absolute standard of rest, the "ether." Suppose also that light travels at velocity c with respect to this ether but gets added and subtracted the way Frisbee velocities do. Then, because the Earth changes direction as it orbits the Sun, at times someone standing on the ether would say the velocity of light was c + v, and six months later c − v. Experiments showed this did not happen.

tory floor. But as the year progressed, the laboratory would turn in space. Just as with the Frisbee on the train, to an observer standing immobile outside the Earth "on the ether," sometimes the light would be traveling in one direction, and, six months later, it would be traveling in the opposite direction. If there was an ether, sometimes the measured velocity of light should be *c plus* the velocity of the Earth and at other times it should be *c minus* the velocity of the Earth.

The results of the experiment were negative. Michelson and Morley found no "ether wind"; the velocity of light appeared to be the same in all directions. Natural philosophers began to emit cries of desperation.

THE FACE IN THE ENCHILADA

The cries were answered only in 1905, the year physicists refer to as the *annus mirabilis*, the year of miracles. In 1905 the first Russian revolution took place; Europe was in crisis over the Morocco question. But what future historians will remember is that Albert Einstein, then a twenty-six-year-old patent clerk in Switzerland, wrote six

WHO'S WHO ☞

Albert Einstein (1879–1955)
The magnitude of his achievements is hard to overemphasize. For this chapter, Einstein is the name you need to know.

scientific papers. Four were epoch-making. In the first he provided the most accurate calculation of Avogadro's number, and by the way introduced the idea of a quantum of light (Chapter 7); for this paper he won the 1921 Nobel prize. Six weeks later he explained Brownian motion (Chapter 2), and hence took one of the final steps in establishing the reality of atoms. Within another two months he submitted a paper containing what is now known as the **special theory of relativity.** Hardly three more months had passed when he chose the wrong letter in deriving a little equation, $L = mc^2$.

It is perhaps difficult for nonscientists to appreciate the magnitude of Einstein's achievement.[1] It was a feat perhaps never before equaled in the history of science, and one that almost certainly never will be again. And yet, his greatest work was still to come.

How did Einstein solve Maxwell's dilemma? With characteristic boldness, he took the bull by the horns. Since no one could detect an ether, he peremptorily declared it null and void. Einstein's decree was formulated in the

Principle of Relativity: There are no absolute refer-

---

[1]The feat is so hard to believe that since 1905 rumors have periodically circulated that Einstein's wife, or someone else, had a large hand in the work. Abraham Pais, a noted authority on Einstein, has spent a year investigating such claims and has found them groundless. See Pais's *Einstein Lived There* (Oxford; 1994).

ence frames and no absolute velocities. All velocities are relative. Maxwell's equations hold without modification in all frames moving at constant velocities.

The first two sentences could have been written by Newton. He had already recognized that his laws hold in all frames moving at constant velocity, a fact discussed above. A train moving at 150 kilometers per hour and a train at rest are no different to the laws. Einstein merely declared that the same must be true for Maxwell's equations.

But how then to explain the curious fact that in Maxwell's equations c, the velocity of light, doesn't get added or subtracted the same way ordinary velocities do? Again, Einstein took the bull by the horns and merely declared that Maxwell was right: c was c and shall always be c. In other words, he postulated

> **The Constancy of the Speed of Light:** The speed of light is not relative. It is always measured to be 3×10^8 meters per second by any observer, regardless of the observer's own velocity with respect to anything else.

This was the revolutionary postulate. We have already convinced you that it contradicts everything Galileo and Newton taught about how to add and subtract velocities. Yet, on the basis of this assumption Einstein constructed a theory that not only explained many observations of the day and put Maxwell's equations in a broader context, but completely changed our notions of space and time. Everyone but Einstein called this theory the special theory of relativity.

Before going into the space and time aspects of relativity, which have captured the imagination of generations,

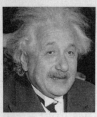

Albert Einstein— His appearance may have been in disarray, but his thoughts certainly were not. To this day, his brilliance in this field remains unparalleled.

we should note that the title of Einstein's paper was "On the Electrodynamics of Moving Bodies"; that is, Einstein was not thinking so much about space and time as about Maxwell's equations.

To get an idea of how Maxwell's equations fell out of Einstein's theory, recall that Coulomb's law gave the electric field around a charge at rest: $E = k_c\, q/r^2$. But relativity deals with moving reference frames. We must therefore ask lots of stupid questions, such as, "What do we mean by 'at rest?'" We mean "at rest with respect to a certain creature A." Creature A will say, "I am at rest, the charge is at rest, I see an electric field." But creature B, zooming by at a high velocity, could say, "I am at rest; that electric charge is moving." Now, as we know, a charge in motion forms an electric current. And according to Ampère's law, an electric current produces a magnetic field. Consequently, creature B will say, "I see an electric current and I feel both an electric and magnetic field."

What were once two distinct Maxwell equations (Coulomb's law and Ampère's law) now turn out to be the same law! The only difference is that one refers to a static charge and the other applies to a moving charge. The same is true of the other two Maxwell equations (the monopole law and Faraday's law). Under relativity, they "fold up" into one equation! Thus, Einstein went one step beyond Maxwell. Not only were the electric and magnetic fields two manifestations of one electromagnetic field, but they weren't even distinct manifestations. It is a bit like Spider-Man. Sometimes he is a superhero, sometimes he is Peter Parker; his wife doesn't know who he is. Which you see depends entirely on your frame of reference.

NEWTON GETS BENT OUT OF SHAPE

The assumption that the speed of light is always a constant had consequences far more reaching than the binding of the electric and magnetic fields into one glittering entity. It required that Newton's laws themselves be altered. For example, we have already mentioned that the usual Galilean and Newtonian rules for adding and subtracting Frisbee velocities don't apply to light. That's not all. We shall see below that the speed of light cannot be exceeded. This means that if creature A on the ground sees creature B in a rocket traveling at one-half the speed of light, and creature B sees a third creature, C, in another rocket traveling at three-quarters the speed of light, creature A *cannot* add the velocities of B and C and say that from the ground creature C appears to be moving at 1¼ the speed of light. The rules of addition change. The Newtonian expressions for mass, force, and energy also get modified in relativity.

All this is because the constancy of the speed of light requires the Newtonian notions of space and time to change. In fact, after 1905, space and time could never be thought of as separate again.

In daily life we tend to think of space and time as distinct and absolute, commuter train delays and Dali's melted watches notwithstanding. Meter sticks measure distances; two good meter sticks will agree when measuring the length of the same garbage can. Clocks measure time; if two firecrackers go off simultaneously at the ends of the street, two good watches will record the same time. Einstein showed that this is all not quite true.

On page 125 we present the famous "light clock," a reasonably comprehensible demonstration of some of these ideas, requiring only basic algebra. It shows that the distances and the times a creature measures depend on the creature's frame of reference. For instance, a creature on a train (t) may measure the time between two sips of wine to be t_t. But a creature watching from the ground (g) will measure that time to be

$$t_g = \frac{t_t}{\sqrt{1 - v^2/c^2}}, \qquad (1)$$

where v is the velocity of the train relative to the ground and c is the speed of light. Equation (1) represents the famous **time dilation**. Notice that as v gets closer and closer to the speed of light, c, the denominator gets closer and closer to zero and t_g gets much larger than t_t. Time "dilates." Also notice that unless v is close to c, t_g is nearly the same as t_t. For this reason relativistic effects are not noticed in daily life. Time dilation is responsible for the infamous "twin paradox," in which one twin zooms off to another galaxy and returns much younger than the twin who stayed behind. All true according to relativity.

The light clock can also be used to show that distances measured on the train will not be the same as distances measured from the ground. For instance, if L_t is the length of the train as measured by creatures on it, its length on the ground will appear to be

$$L_g = \frac{L_t}{\sqrt{1 - v^2/c^2}} \qquad (2)$$

As v gets close to c, L_g gets close to zero! Here we have the famous "Lorentz contraction," which shows that the sizes of moving objects appear to shrink.

Yet another result of relativity is that the mass of an object gets larger as its velocity increases:

$$m = \frac{m_o}{\sqrt{1 - v^2/c^2}}. \qquad (3)$$

Here m_o represents the mass of an object at rest, its "rest mass." If you imagine accelerating your creature by pushing on it, you see that you require a greater and

DEMO 1

The Light Clock !!!

For this demo you need relationships among distance, time, and velocity, and the Pythagorean theorem, and you need to refer carefully to the figures.

Any device that executes a regular, repetitive motion can be used as a clock. Consider, then, a clock made of two mirrors, separated by a distance L_t. (The subscript t will always stand for train.) A Ping-Pong ball bounces between the mirrors. Each round trip—a distance $2L_t$—represents one tick. The whole clock is mounted on a train, as in figure **a**.

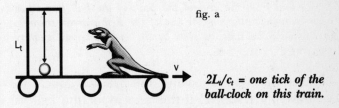

fig. a

$2L_t/c_t$ = *one tick of the ball-clock on this train.*

Assume the ball travels up and down with velocity c_t. Because time = distance/velocity, to make one round trip $2L_t$ requires a time

$$t_t = 2L_t/c_t. \qquad (1)$$

This is the time for one tick as measured by a creature standing on the train.

Suppose also that the train is moving with a velocity v relative to the ground. A creature on the ground will see the ball travel in the path shown in figure **b**.

DEMO 1

(continued)

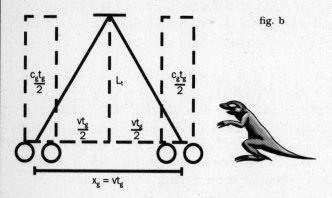

fig. b

Notice that from the ground the path of the ball forms two right triangles, allowing the use of the Pythagorean theorem to calculate the time as seen by the ground-based creature (see text).

Because the train is moving, the ball will appear to be traveling faster. Call the velocity of the ball as seen from the ground c_g and the time for a tick as measured on the ground t_g. From the figure, the Pythagorean theorem gives:

$$c_g^2 t_g^2 / 4 = L_t^2 + v^2 t_g^2 / 4.$$

Subtracting $v^2 t_g^2 / 4$ from both sides and factoring out the t_g gives:

$$t_g^2 (c_g^2 - v^2) / 4 = L_t^2$$

or

$$t_g^2 = 4 L_t^2 / (c_g^2 - v^2). \qquad (2)$$

DEMO 1

(continued)

But from (1) $L_t = c_t t_t/2$. Squaring this and plugging it into (2) gives:

$$t_g^2 = c_t^2 t_t^2 / (c_g^2 - v^2). \qquad (3)$$

This expression gives the time as measured on the ground compared with that measured on the train.

Equation (3) looks complicated, but examine figure **c**.

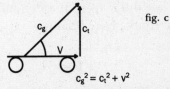

fig. c

$$c_g^2 = c_t^2 + v^2$$

If the velocity of the ball and train get added according to the rules of Galileo, then you also need the Pythagorean theorem to calculate the velocity of the ball as seen from the ground (c_g).

According to Galileo's rules for adding velocities, the velocity of the ball as seen from the ground will be $c_g^2 = c_t^2 + v^2$. (Here we used the Pythagorean to add velocities.) If you substitute this expression for c_g^2 into the denominator of (3), everything cancels and you get nothing but

$$t_g = t_t. \qquad (4)$$

That is, as you expect from daily life, a clock read from the ground or from a train tells the same time.

Well, that isn't too exciting. But this dull result was due to the assumption that velocities got added according to the rules of Galileo, as in figure **c**. Let's now assume that the Ping-Pong ball is made out of light.

DEMO 1

(continued)

According to Einstein, the speed of light is always the same, regardless of who measures it. Thus the velocity of light *does not* get added as in figure **c**. Instead $c_g = c_t$. Let's just call both c. Then (3) becomes

$$t_g^2 = c^2 t_t^2 / (c^2 - v^2). \qquad (5)$$

Factoring out the c^2 from the denominator and taking square roots gives, finally,

$$t_g = \frac{t_t}{\sqrt{1 - v^2/c^2}} \qquad (6)$$

as stated in Equation (1) of the text. This is the famous **time dilation** of relativity. The time measured by two observers is not absolute, but depends on their frame of reference.

We can also use the light clock to verify that the **spacetime distance** is invariant (see text). That is, we wish to check whether

$$x_g^2 - c^2 t_g^2 = x_t^2 - c^2 t_t^2 \qquad (7)$$

where x is the spatial distance between two events (say ticks on a clock or sips of wine) and t is the time between two events. Refer again to figure **b.** To a creature on the train, the ticks of the clock always occur in the same **place.** So there is no spatial distance between ticks and $x_t = 0$. A creature on the ground, however, sees the train move between ticks, so x_g is not zero. Indeed, if the train is moving

DEMO 1

(continued)

at velocity v, then $x_g = vt_g$. Thus (7) becomes

$$v^2 - c^2t_g^2 = -c^2t_l^2$$

or

$$t_g^2(v^2 - c^2) = c^2t^2.$$

But this is the same as

$$t_g = \frac{t_l}{\sqrt{1 - v^2/c^2}},$$

which is true by (6). We have thus verified the **invariance of the spacetime distance** for the light clock, but it is in fact always true.

greater force to push the creature yet faster, because its mass is always growing. When $v = c$, the creature's mass is infinite, so you require an infinite force (or energy) to get it going any faster. For this reason Einstein declared that nothing can travel faster than the speed of light. Indeed, we may elevate this statement to a law of nature:

> **The Speed Limit:** No material object may travel faster than the speed of light.

Many infidels think that the Speed Limit is something Einstein just dreamed up one day while smoking his pipe and that eventually we will break it. After all, where would science fiction be without faster-than-light travel? Sorry, science fiction is one thing, the real world is another. The universe has laws. Behold, this is one of them.

Finally, for laycreatures, relativity's most famous result is contained in the equation $E = mc^2$. Regardless of how many times it has appeared on blackboards in *New Yorker*

cartoons, it is truly a great result. But it is not as mysterious as you might think, and using the knowledge you have absorbed so far plus one result from Chapter 7, we can actually derive it. $E = mc^2$ tells us that there is an equivalence between mass and energy, that every mass has a certain amount of energy associated with it and vice versa. The c^2 is the conversion factor between them. This equation does not tell you how to get the energy out of matter, nor whether it is even possible. It just says that every mass has an associated energy. Normal chemical reactions release less than a millionth of mc^2; a hydrogen bomb releases less than one percent. In any case, the equivalence of mass and energy requires that the law of conservation of matter be modified to

> **The Conservation of Mass-Energy:** The total amount of mass-energy in an isolated system does not change.

LIVING WITH RELATIVES

So far we have discovered that the electric and magnetic fields are two aspects of the same thing, and that energy and mass are as well. But it is not evident from the equations we have written down how space and time get tied together in relativity. (The author decided to keep you in suspense.) It's actually not too surprising if you think about it. Consider a creature sipping wine on a train. To that creature the two sips take place at the same *place* but different *times*, maybe a few seconds apart. But to a creature watching from the ground, the two sips take place at both different *times* and different *places*. So what is just "time" to one creature is both "space and time" to another.

Relativity shows precisely how the two are related. And although the two creatures can't agree on whether the sips took place in "time" or in "time and space," that doesn't mean they can't agree on anything. From the light-clock demonstration, one can get another result, that a certain quantity is *the same* whether measured on the train or on the ground. This quantity is

$$x_g^2 - (ct_g)^2 = x_t^2 - (ct_t)^2 \qquad (5)$$

where again the t_t and t_g represent the times between the two sips of wine as measured on the train and on the ground, and x_t and x_g represent the spatial distance between the sips. A quantity that is the same in all reference frames is said to be **invariant**.

If you stare at (5) long enough to see that either side looks a bit like the Pythagorean theorem for right triangles: $a^2 + b^2 = c^2$ (c being the length of the hypotenuse and a and b being the lengths of the other two sides). The figure below details how the Pythagorean theorem is used to find the distances between objects. The important point is that the distance between your house and the nearest Dunkin' Donuts is also an invariant. Faced with a doughnut attack, you might pace off the distance, s, directly "as the crow flies," and find it is five blocks. But usually houses in between foil such a straightforward attempt. You may then resort to walking three blocks north and four blocks east. Computing the diagonal with the Pythagorean theorem gives $s^2 = 3^2 + 4^2$. Once again, $s = 5$. You may resort to more complicated strategies, starting ten blocks west, for example. But no matter how you do it, you will always find that five blocks separate your house from Dunkin' Donuts.

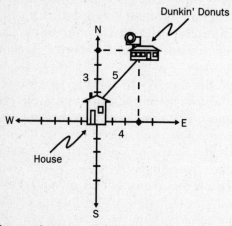

*The distance from your house to Dunkin' Donuts remains the same no matter how you get there. It is **invariant**.*

In relativity, a quantity usually called $d\tau^2 = x^2 - (ct)^2$ plays the same role as distance does in Euclidean geometry (τ is the greek letter tau). But in this case, spatial distances are mixed up with differences in time. For this reason, we refer to this new quantity $d\tau$ as the **spacetime** separation between two events. A creature on the train sees the sips of wine take place at different times but at the same place. A creature on the ground disagrees on the time lapsed between the sips and also sees them take place at different places. But one thing they both agree on is that the spacetime separation between the sips is the same.

We see that in relativity not all things are relative. The speed of light is the same for everyone. So is the spacetime distance $d\tau$. Other quantities are also invariant, for instance $E^2 - B^2$, where E and B are the strengths of the electric and magnetic fields. Neither does $E^2 - p^2$ change, where this time E is the energy and p is the momentum of a particle. The idea that "everything is relative" has had an enormous impact on art and culture. You may wish to try it out at a cocktail party, but beware the inappropriate metaphor. Indeed, Einstein himself preferred not to call relativity relativity. He was more concerned with what remained the same when viewed from different reference frames and hence called his theory the "theory of invariants." Add one more topic the theory of relativity can be applied to: its own name.

ESOTERIC TERMS
(es-ə-'ter-ik tərms)

- *Ether*—Doesn't exist. Was supposed to be the medium that carried light waves in the same way air carries sound waves.

- *Time dilation*—This inexact term refers to the fact that to a stationary observer moving clocks seemingly run slow.

- *Lorentz contraction*—The apparent shrinking of the lengths of moving objects when viewed by stationary observers. (Could also be used to describe the effect of a moving object on Mrs. Lorentz during her last hours of pregnancy.)

- *Invariant*—Noun or adjective. As noun, refers to any quantity that is the same to observers in different frames of references.

- *Spacetime*—The arena in which events occur in relativity theory. In Einstein's scheme, space and time are no longer separate but are combined into one entity: spacetime.

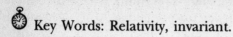

SUMMARY

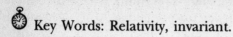 Key Words: Relativity, invariant.

 Key Definition: An invariant is something that doesn't change from one frame of reference to another.

 Key 'cepts: All reference frames moving at constant velocities behave the same, as if they are at rest. All laws of physics are the same in any of these frames. The assumption that the speed of light be invariant requires that Newtonian concepts be modified for high velocities. Many other quantities in relativity are also invariant.

CHAPTER 6

INSIDE THE NUCLEUS

CURIES AND CURIOUSER

YOU MUST REMEMBER THIS

Physics has had great impact on everyday life. This is nowhere clearer than in the study of nuclear physics.

THE GUIDING LIGHT

In our continuing saga, atoms had almost become real by the close of Episode Two. Young Albert Einstein, forced to take refuge in a patent office, calculated the best value for Avogadro's number. Six weeks later he explains Brownian motion as the collision of molecules with pollen. But will the world accept atoms? What other evidence is there? Will Thomson discover X rays before Roentgen? Will Pierre Curie die of radiation sickness before being run over by a coach? For the answer to these and other exciting questions, stay tuned for Episode Six of *Physics and Destiny*.

After that commercial interruption we may return to sanity. By now you probably realize that science does not develop along textbook lines. Nowhere is this more apparent than in the birth of atomic and nuclear physics. Einstein in 1905 took two of the last steps in establishing the existence of atoms, but as far as he was concerned they were structureless little billiard balls, no different from the atoms Dalton had proposed a century earlier. Yet, even in the last decade of the nineteenth century, not only was there compelling evidence that atoms existed, but physicists were starting to discover hints that the atom was not "indivisible."

The onslaught began with a keen eye. A favorite pastime among physicists at the end of the century was to amuse themselves with so-called "Crookes tubes," named after their inventor, Sir William Crookes. Crookes tubes[1] were nothing more than sealed glass tubes from which most of the air had been evacuated and into which electrodes (flat pieces of metal) had been inserted at each end. When a high voltage was placed between

---

[1] Known as "Geissler tubes" on the Continent, after Johann Geissler, one of the other inventors.

the cathode (**negative electrode**) and the anode (**positive electrode**), the tube would light up. Sometimes a metal object was inserted between the electrodes; if so, its shadow would be cast against the anode end of the tube by the "cathode rays" emitted from the cathode. One almost wishes that Crookes tubes had never caught on; each day their descendants emit mind-disintegrating rays in America's 100 million television sets.

WHO'S WHO

Wilhelm Roentgen (1845–1923)
Accidentally discovers X rays and makes

J. J. Thomson (1856–1940)
jealous. But Thomson gets his revenge by discovering the electron.

Henri Becquerel (1852–1908)
is also stunned by Roentgen's discovery and so discovers radioactivity. Little do his colleagues

Marie (1867–1934) and **Pierre** (1859–1906) **Curie**
realize how dangerous it is, and they go on to discover polonium and radium, not knowing that Marie's death is hastened by radiation sickness. But Pierre never finds out because he is run over by a coach. At about the same time,

Ernest Rutherford (1871–1937)
falls in love with the atom, and discovers the atomic nucleus, but not before he and

Frederick Soddy (1877–1956)
discover the transmutation of the elements by radioactivity and come up with the idea of half-

**WHO'S
H
O
☞**

(continued)

life. Determined not to let Rutherford and Soddy corner all the Nobel prizes,

Henry Moseley (1887–1915)
establishes the concept of atomic number, the number of charges in the atomic nucleus, and rewrites the periodic table. Destined for glory, he marches off to Gallipoli and is killed. Grieving, Rutherford takes

James Chadwick (1891–1974)
under his wing. Together they discover the proton, then Chadwick goes on to discover the neutron. Shortly after,

Frédéric (1900–1958) and **Irène** (1897–1956) **Joliot-Curie**
bombard elements with alpha particles and find they become radioactive. Their discovery of artificial radioactivity really pisses off

Enrico Fermi (1901–1954)
who doesn't like the French and English getting all the credit, so he begins bombarding atoms with neutrons and converting them into other elements. He later takes out his frustrations on the atomic bomb, but only after

Otto Hahn (1879–1968) and **Fritz Strassman** (1902–1980)
extend Fermi's work and discover nuclear fission, making it possible for everyone to blow themselves to bits.

In 1895 Wilhelm Roentgen noticed that the glass walls of such a discharge tube (as we now call them) glowed when bombarded with cathode rays. He covered the tube with black cardboard but then caught sight of a fluorescent screen glowing on a nearby workbench. Whatever the cause of the screen's strange behavior, Roentgen knew it was not cathode rays; they did not travel through more than a few centimeters of air. He held his hand between the tube and the screen and found that it still glowed, but that he could see the shadows of his bones. Further investigation by Roentgen showed that this new radiation penetrated many materials, could expose photographic plates, was not deflected by electric or magnetic fields, and did not seem to undergo interference in a Young's experiment (p. 107). Roentgen named his mysterious new radiation **X rays.** For the discovery he won the first Nobel prize in physics, in 1901.

Nobel prizes come and go, but the discovery of X rays stunned the world. Within a few months doctors began using X rays for medical diagnosis. Not realizing how dangerous they were, several died from overexposure. Thus atomic physics and the public's fear of it were born at the same moment.

The exact nature of X rays remained a mystery for fifteen years. It will not give away too much of the punch line to say that in 1912 Max von Laue suggested that X rays were merely light rays of extremely short wavelengths—so short that Roentgen could not observe the usual interference effects. He reasoned that atoms in crystals were close enough together that they might serve as the slits in a Young's experiment (Chapter 4). His idea was confirmed by several experiments, in particular those of William Henry and William Lawrence Bragg. A rash of Nobel prizes ensued. X rays are indeed electromagnetic waves, whose wavelength is more or less one thousand times shorter than visible light. They are produced when **cathode rays** strike metal and can be damaging to your health.

DAYS OF ELECTRONS

To return to 1895. Not everyone died of X rays. J. J. Thomson had been studying cathode rays when Roentgen made his announcement. He undoubtedly felt the pang of missed opportunity, for the pace of his research quickened. Within two years Thomson noticed that cathode rays could be deflected by both electric and magnetic fields. That meant that the rays consisted of charged particles; in fact, Thomson determined that they consisted of "negative corpuscles," i.e., negatively charged particles. By an ingenious experiment that physics undergraduates reproduce today, he measured the "q/m ratio" of the corpuscle, or its "charge-to-mass ratio." The fact that this ratio was not zero or infinity meant that the particle had a definite charge and a definite mass, though Thomson's experiment could not give them individually.

Now, Thomson had also measured the q/m ratio for **hydrogen ions.** Hydrogen ions were particles that had all the same properties as hydrogen atoms except that, while an electric field did not deflect atoms, it deflected the ions oppositely from the "negative corpuscles." This meant hydrogen ions were positively charged. Also, the q/m ratio of the negative particles seemed to be about 1,000 times *larger* than the q/m ratio of the hydrogen ions. Assuming that the charges on each were the same, the new particle must be 1,000 times *lighter* than hydrogen. The conclusion seemed nearly inescapable: the atom—whatever *that* was—was no longer the smallest entity. "Negative corpuscles" must be part of the atom. Despite the uncertainties, Thomson had indeed discovered the first subatomic particle, which soon became known as (add trumpet sounds . . .) the **electron.**[2]

THE YOUNG AND THE RADIOACTIVE

Henri Becquerel of Paris was another physicist stimulated by Roentgen's investigations. When he heard a report of Roentgen's work, Becquerel immediately decided to try to find other materials that gave off X rays. Over the next ten days he tried various substances without success. Then he picked an ore containing uranium. He wrapped up a photographic plate in black paper and sprinkled the "uranium salt" (as it was known) onto the paper. He exposed the whole package to the Sun. Upon developing the plate he saw the silhouettes of the salt grains in black on the negative. Becquerel mistakenly thought that the sunlight had "activated" the crystals,

[2]For more on these and other developments in early twentieth-century physics, see Aaron J. Ihde, *The Development of Modern Chemistry* (Dover, 1984).

DEMO 1

Thomson's Experiment !!

For this demo you need all the relations among acceleration, distance, and velocity in Demo 1, Chapter 1, as well as the expressions for electric and magnetic force from Chapter 4.

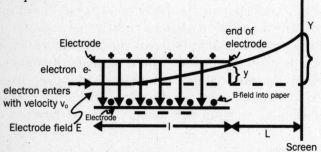

Electrode

electron e-

electron enters with velocity v_0

Electrode

Electrode field E

end of electrode

y

B-field into paper

l

Y

L

Screen

Schematic Diagram of Thomson's Experiment

Like Bernoulli's derivation of Boyle's law, Thomson's experiment to measure the q/m ratio of the electron is a good illustration because it incorporates much of what we have learned. The experimental setup is shown in the figure. A beam of electrons traveling horizontally at velocity $v_.$ is passed through an electric field, E. From Chapter 4, the force exerted by the field on a charged particle is simply $F = qE$. Since $F = ma$ by Newton's second law, we have

$$a = F/m = qE/m. \quad (1)$$

From Equation 8 on page 31, this force will deflect the electron a vertical distance y in a time t given by

$$y = (\tfrac{1}{2})\, at^2 \quad (2)$$

DEMO 1

(continued)

or, using a from (1)

$$y = (\tfrac{1}{2})qEt^2/m. \qquad (3)$$

We need to get rid of t, since it's too small to measure. Because there is no horizontal force, the horizontal velocity v_0 never changes. (Newton's first law: no force, no change in velocity.) That means the time it takes to travel the length of the electrodes, the distance l, is just $t = l/v_0$. Then (3) becomes

$$y = (\tfrac{1}{2})qEl^2/mv_0^2. \qquad (4)$$

This is the amount of vertical deflection after the electron has crossed the horizontal distance l. The trouble is, we don't know v_0. Here is where Thomson got clever. From Chapter 4, Equation 4, if one adds a magnetic field at right angles to the electric field, the total force on the electron is

$$F = qE - qv_0B \qquad (5)$$

where the minus sign indicates the direction of the B-field.

Now, Thomson first turned on the electric field, so that the electron got deflected. Then he switched on the magnetic field and adjusted its strength so that the electron's path became undeflected again. That meant F was zero. In which case (5) gives:

$$v_0 = E/B. \qquad (6)$$

Plugging (6) into (4) yields

$$y = (\tfrac{1}{2})qB^2l^2/mE \qquad (7)$$

DEMO 1

(continued)

or

$$q/m = 2yE/B^2 \qquad (8)$$

for the q/m ratio. Note that everything on the right-hand side of (8) can be measured. Thus q/m can be determined.

    You can stop now if you want. On the other hand, you might object that you only measure the deflection once the particle strikes the screen at Y, not y. True, but after the electron reaches l, the end of the electrodes, it no longer experiences any force, so it travels in a straight line at constant velocity until it clobbers the screen at Y. Thus we have simply $Y = y + v_v t_L$, where v_v is the *vertical* velocity the electron has at l and t_L is the time it takes to traverse the distance L.

    So to get Y we need v_v and t_L. The latter is merely $t_L = L/v_o$. The vertical velocity at l is just $v_v = at$; we know a from (1) and as before, $t = l/v_o$. To express q/m in terms of Y instead of y requires only a few more steps, which we leave to the reader.

causing them to emit X rays, which penetrated the paper and exposed the film.

What happened next became legendary. As a good scientist he had planned to repeat the experiment—but the sky was overcast. He put the wrapped-up plate with uranium on top in a drawer for several days, then for some reason decided to develop it anyway, expecting perhaps only the faintest images of the salt crystals. To his amazement, the silhouettes were even more distinct than when he had exposed the package to sunlight! Whatever was going on had nothing to do with the Sun. The uranium ore was emitting something by itself.

It also had nothing to do with X rays. Further experiments by Becquerel showed that his "uranic rays" were not the same as Roentgen's rays, although both remained mysterious. Any uranium ore gave off uranic rays, and months in a dark drawer made no difference to their intensity. By 1898 two of Becquerel's colleagues, Marie and Pierre Curie, isolated from one ton of a mineral called pitchblende a gram of an active ingredient that was emitting uranic rays. Marie christened the new element polonium, after her homeland, and named the spontaneous emission of uranic rays (you guessed it) **radioactivity**.

The nature of the uranic rays was clarified largely through the work of J. J. Thomson's student Ernest Rutherford. In studying the radioactive elements uranium and thorium, he decided they gave off at least two types of radiation; he named them **alpha** and **beta.** Both alpha and beta rays could be deflected by electric and magnetic fields, but the alphas could be stopped by metal foils .003 centimeters thick, whereas the betas required a foil about 100 times thicker. In France a third radiation was soon recognized, the **gammas,** which were undeflected by electric or magnetic fields and which could be stopped only by about a centimeter of lead.

By 1902 Rutherford demonstrated that beta rays had all the same properties as Thomson's electrons. He thus declared them the same and the identification has stood the test of time. The term **beta particle** is still often encountered today; it is merely an antique synonym for **electron.** We can also reveal, without lessening too much of the suspense, that gamma rays turned out to be similar to X rays. That is, gamma rays are merely light, but of ultra-short wavelengths, even shorter than X rays. Like

X rays, gamma rays can be extremely hazardous to your health. So, now we know: radioactive elements emitted electrons, light, and the puzzling alphas.

ONE HALF-LIFE TO LIVE

During the same years, Rutherford was carrying on a series of important experiments with his colleague Frederick Soddy. In 1900 Rutherford had reported that thorium gave off a radioactive gas. The Curies had also reported that the element radium (which they had isolated from uranium ore in 1898) gave off a radioactive gas. Together, Soddy and Rutherford identified the gas as **argon.** The experiment, wrote Soddy, "conveyed the tremendous and inevitable conclusion that thorium was slowly and spontaneously transmuting itself into argon gas!"[3] Soddy and Rutherford had observed the formation of one element from the disintegration of another. Elements could be transmuted after all. The alchemists had been right.

Over the next few years, Soddy and Rutherford traced the complicated pathways by which the various radioactive elements decay, all the while emitting alpha, beta, and gamma radiation. In doing so, they introduced the concept of **half-life,** which is the amount of time it takes for half of the atoms of a given element to decay. Half-lives vary widely. Soddy and Rutherford found, for example, the half-life of uranium to be 4.5 billion years, radium's to be 1,620 years. Today their discoveries form the basis of **radioactive dating.** (In fact, radioactive dating often comprises the sum of many a physicist's social life.)

In any case, the implications of radioactivity for civili-

---

[3]Richard Rhodes, *The Making of the Atomic Bomb* (New York: Simon and Schuster, 1986), p. 43.

DEMO 2

The Age of the Earth !

This demo requires you to understand half-life.

Radioactive isotopes have been regularly used to date the age of the Earth. To illustrate the method, the most prevalent type of uranium is the isotope ^{238}U, which gradually decays to lead with a half-life of 4.5 billion years. The more radioactive isotope of lead, ^{235}U, decays with a half-life of 700 million years. Typically in mineral deposits the ratio of ^{235}U to ^{238}U is .7 percent, or .007. Let us assume the Earth was formed with equal amounts of these isotopes, meaning their initial ratio was one to one. If so, how old is the Earth?

This problem is simple enough that the answer can be estimated in a stupid way. Imagine a race between ^{235}U and ^{238}U. They start off together in the same amount, one unit apiece. By definition, half-life is the time it takes for half of an element to decay. That means after 700 million years, the amount of ^{235}U has halved to .5. After another 700 million years it has halved again to .25. After seven half-lives it has decreased seven factors of 2, to 2^{-7}, or .008. But seven half-lives of ^{235}U is about one half-life of ^{238}U. Thus, in this time the amount of ^{238}U has decreased to .5. Their ratio at that point is then .008/.5 or .016. But after another 700 million years the amount of ^{235}U will have halved again, bringing the ratio down to .008, about what is actually observed. Thus the age of the Earth should be eight half-lives of ^{235}U, or 5.6 billion years. A more exact calculation gives a similar answer. The Earth was not created in 4004 B.C.

zation were not lost on Rutherford and Soddy. They noted that the energy given off by one gram of radium was between 100 million and 10 billion calories (a 22-year diet) and was therefore up to "a million times as great as the energy of any molecular change." In 1904 Soddy speculated that if the energy bound up in the atom "could be tapped and controlled, what an agent it would be in shaping the world's destiny! The man who puts his hand on the lever . . . would possess a weapon by which he could destroy the earth if he chose." All this before $E = mc^2$.[4]

JELLIUM HOSPITAL

Although Rutherford and Soddy were measuring the radioactive decay of elements and speculating on the atomic bomb, they did not know what atoms were made of. After Thomson's discovery, it was clear that atoms must contain electrons. At the same time it became clear that when electrons were stripped away from atoms—a process called **ionization**—they left a positively charged **ion,** which contained virtually all the mass of the entire atom. But in order to keep atoms neutral, their natural state, the amount of positive charge on the ion must equal the amount of negative charge on the electron. Thomson himself thought an atom consisted of electrons embedded in some nebulous jelly of "positive electrification." His model, for obvious reasons, became known as the "plum pudding" model, or "jellium."

The key to the correct model lay in alpha radiation. A good rule in physics is that if a tool works, use it until it breaks. So Rutherford made q/m(charge-to-mass ratio) measurements of alpha particles. By 1906 he had deter-

---

.[4]Ibid., p. 49.

mined that the q/m value for alphas was *one-half* that of hydrogen. Now, helium ions are *four* times as heavy as hydrogen ions. If they had *twice* the charge, their q/m ratio would be one-half hydrogen's. For this reason Rutherford suspected alphas were nothing other than helium ions, a suspicion supported by Soddy's observation that helium was liberated in radioactive decay.

The guess was confirmed by Rutherford and Thomas Royds in 1908 by an ingenious experiment. They knew that the newly discovered gas radon gave off alpha particles. They placed radon in a glass bulb with walls so thin that alphas passed through them into an outer tube, from which the air had been evacuated. There the alphas collected for a few days. The experimenters then sent an electric current through the alpha-gas and it lit up like a neon sign.

Or in this case a helium sign; the colors it gave off were exactly those of helium. The pattern of colors—the **spectrum**—of light given off by a hot gas is its fingerprint: no two gases produce the same spectrum. For this reason, spectra have been an unsurpassed diagnostic tool in physics. As to *why* gases emit the exact colors they do, that was one of the most important questions of physics in the early twentieth century and we will hear more about it in Chapter 7. For the moment we can say, almost without exaggeration, that spectra have turned out to be the key to the universe; using this key, Rutherford confirmed that **alpha particles were ionized helium atoms.**

Rutherford put his alphas to work. In 1908 he directed two colleagues, Hans Geiger (yes, the inventor of the counter) and Ernest Marsden, to bombard an incredibly thin gold foil with alpha particles. You must understand this foil was only .00006 centimeters thick, 50 times thinner than the .003 centimeters usually needed to stop alphas. Most of the alphas passed straight through, but sometimes they would get deflected through a small angle. And occasionally—just occasionally—an alpha would bounce straight back in the direction it came. Years later Rutherford described the event in a statement so famous its inclusion in every physics book is obligatory: "It was quite the most incredible event that has ever happened to me in my life. It was almost as incredible as if you fired a 15-inch shell at a piece of tissue paper and it came back and hit you."[5]

The results confirmed in a spectacular way that, as others had suspected, the atom was mostly empty space. In order to get the backward bounces, a head-on collision was necessary: most of the atom's mass had to be concentrated in an area about 10^{-14} meters in diameter! Rutherford had discovered the atomic nucleus.

ALL MY PARTICLES

Following this historic experiment, Rutherford formulated the "solar system" picture of the atom we often rely on today: a tiny nucleus 10^{-14} meters in diameter, surrounded by electrons orbiting at a distance 10,000 times larger. Thus if the nucleus were the size of Saddam Hussein's head, the electrons would be orbiting somewhere around the outskirts of Baghdad. Indeed, the atom was mostly empty space.

However, at the time, no one was sure what the nucleus consisted of. Because like charges repel and the positively charged alphas bounced off the nucleus, it was clear the nucleus consisted of positive charges. But how much? From Demo 3 on page 151, you see that the incoming alpha particles in Geiger and Marsden's experiment were deflected by an amount that depended on

---

.[5]Ibid., p. 49.

DEMO 3

The Size of the Nucleus!!

This demo requires the definition of kinetic energy from Chapter 3, Coulomb's law from Chapter 4, and the precise definition of force from the Talmud in Chapter 1.

The size of the atomic nucleus is easy to estimate with a few concepts already in hand. Rutherford's colleagues fired alpha particles at a piece of gold foil. The kinetic energy of these alphas was about 10^{-12} joules. Thus, from the formula for kinetic energy (Equation 5 on p. 75),

$$(½)mv^2 = 10^{-12} \text{ J.} \qquad (1)$$

Now, the force exerted on an alpha particle by the nucleus of the atom is given by Coulomb's law, Equation 1 in Chapter 4:

$$F = k_c q_1 q_2 / r^2. \qquad (2)$$

We will assume that the alphas that bounced directly backward experienced the largest possible force, which corresponds to the smallest possible r. Presumably this is roughly the size of the nucleus. Our task is merely to compute the smallest r.

Just as in Bernoulli's derivation of Boyle's law (p. 76), we use the strict definition of force: change of momentum per unit time. If the momentum of an alpha is $p = mv$ and it bounces backward, its change of momentum is $\Delta p = 2mv$. How long does this take? Most of the action must take place in the vicinity of the nucleus, within a distance of about r. (It's the only distance we have.) The time

DEMO 3

(continued)

it takes to traverse this distance is simply $\Delta t = r/v$. So, roughly,

$$F = (\Delta p)/(\Delta t) = 2mv/(r/v) = 2mv^2/r = k_c q_1 q_2/r^2.$$

Dividing each side by 4 and multiplying by r gives

$$(\frac{1}{2})mv^2 = k_c q_1 q_2/4r. \qquad (3)$$

Notice the left-hand side is just the kinetic energy of the alphas, 10^{-12} Joules. All that is necessary now is to solve for r and plug in the numbers. The constant $k_c = 9 \times 10^9$. The charge on an electron is 1.6 $\times 10^{-19}$ coulombs. An alpha particle has twice this charge and a gold atom about 80 times (Rutherford would not have known this number). Thus

$$r = (9 \times 10^9) \times 2 \times 80 \times (1.6 \times 10^{-19})^2/(4 \times 10^{-12})$$

Working this out yields approximately 10^{-14} meters for the radius of the gold nucleus. This is about 10,000 times smaller than the atom itself.

the nuclear charge. So you might guess that by examining the deflection of the particles you can determine the size of the charge.

A reasonable guess. Geiger and Marsden tried the scattering experiments with materials other than gold. What did they find? Call the size of an electron's charge 1. The size of the nuclear charge relative to the electron will be Z. Call the **atomic weight** (in other words, atomic mass) of hydrogen 1, helium 4 . . . gold 197. They argued from the data that Z, the nuclear charge, always seemed to be roughly one-half the atomic weight. For instance, for gold Z was 80, i.e., 80 electrons' worth of positive charge in the gold nucleus.

Now, in addition to atomic weight, physicists labeled elements by their **atomic number,** which was merely the number that marked the element's position on the periodic table. Hydrogen was 1; helium, 2, and so on. Coincidentally, for most elements, the atomic number was about half the atomic weight. Indeed. **This suggested that Z represented both the atomic number and the number of positive charges in the nucleus.**

The idea was confirmed by Henry Moseley, who in 1913 bombarded various metals with electrons and measured the frequency of the X rays given off. For reasons which will become clear in Chapter 7, he found that the frequency was proportional to Z^2, the square of the number of nuclear charges, and that Z increased by one from each element to the next (in contrast to atomic weights; e.g., hydrogen to helium is a jump of three in weight). In other words, Z was shown by Moseley to be the atomic number.

Moseley's work was largely responsible for the periodic table as we know it today. Before his experiments, periodic tables had always been arranged in order of increasing **atomic weight.** But Moseley found this did not always correspond to increasing charge Z. (For instance, nickel was placed before cobalt because nickel had the lower atomic weight. However, according to Moseley's data, nickel had the higher Z.) To identify Z with the atomic number, then, required reversing the order of several elements. This, in fact, helped make their **periodic properties** (Chapter 2) more consistent and led to the prediction of new elements. A year after publishing his famous results, Moseley was killed in the battle of Gallipoli.

The next question was whether the positive charge in the nucleus was connected with a particle. That question also had to wait for the conclusion of World War I. In

"It says, 'You may already be a Nobel Prize winner.' "

1919, Rutherford and James Chadwick bombarded many elements with alphas. Unknown particles were given off and sent through a magnetic field. By the direction of deflection they were found to be positively charged. Rutherford and Chadwick had identified the **proton,** the positively charged particle in the atomic nucleus. **The size of the proton's charge is the same as the size of the electron's charge, but of opposite sign. Therefore the atomic number Z of an element corresponds to the number of protons in a nucleus.**

AS THE NEUTRON TURNS

The trouble was, when the proton's mass was measured, essentially by determining the q/m (change-to-mass) ratio, it accounted for only about half the total mass of the atom. Something else was required. That something was the **neutron,** discovered by Chadwick in 1932. As its name implies, the neutron is electrically neutral, and so it is undeflected by either electric or magnetic fields. For this reason it proved exceedingly difficult to discover. We content ourselves with stating that the neutron is slightly heavier than the proton and both are about 1,800 times heavier than the electron. **The atomic num-**

ber of an atom is the number of its protons, while the atomic weight of an atom is the number of neutrons plus protons.

Not quite. If you carefully add up the mass of all the protons and neutrons in an atom and compare the total with the atomic weight of the element on a periodic table, you will discover a curious fact. For elements lighter than iron, the atomic weight is a little *smaller* than the mass of the component protons and neutrons, whereas for elements heavier than iron, the atomic weight is a little *larger* than the total mass of the neutrons plus protons.

"Well then," you ask, "where is this extra mass in the heavy elements coming from?" It comes from $E = mc^2$. This is the so-called **binding energy** of the atom, the energy needed to bind the protons and neutrons together into the nucleus. The fact that heavy elements are heavier than the sum of their parts means that they have energy to spare. Elements like uranium have a lot of energy to spare. It is this energy that appears in alpha, beta, and gamma rays when the elements undergo radioactive decay.

The binding energy actually available for consumption depends very much on the particular **isotope** of the chemical element you are considering. Two isotopes (read "versions") of an element have the same number of protons but a different number of neutrons; ^{238}U is the most common isotope of uranium. The 238 represents the atomic weight. Since the atomic number is 92, the number of neutrons is $238 - 92 = 136$. On the other hand ^{235}U has only 133 neutrons; ^{235}U has so much extra binding energy that it is highly unstable.

In the 1930s, the work of Frédéric and Irène Joliot-Curie (Marie's daughter) and Enrico Fermi established that one isotope could be transmuted into another by bombarding it with alpha particles or neutrons. In 1939,

Otto Hahn and Fritz Strassman discovered that under neutron bombardment uranium produces barium and krypton, each with about *half* the atomic weight of uranium. Soon thereafter, Otto Frisch and his aunt Lisa Meitner suggested that under neutron bombardment uranium was undergoing **fission** and releasing its binding energy. You know the rest of the story.

However, not every tale in physics need be dark. For isotopes lighter than iron, the mass (energy) of the component parts is greater than the mass of the end product. So by combining two light isotopes into a heavier one you release the extra energy. Such a process is termed **fusion.** In the Sun, hydrogen nuclei fuse to form helium. The process releases their binding energy. It is this energy that ultimately sustains life on the Earth. Physicists are currently within reach of harnessing fusion energy in machines known as tokamaks. The fuel, the isotope of hydrogen called deuterium (one proton, one neutron), is readily available in seawater. When fusion becomes available, it will provide an energy source without meltdowns or greenhouse gases, and one that will last longer than civilization will. Not a bad result of nuclear physics.

ESOTERIC TERMS
(es-ə-'ter-ik tərms)

- *X rays*—Light wave with a frequency 100 to 1,000 times higher than visible light.

- *Gamma rays*—Light waves with frequencies higher than X rays.

- *Alpha particles*—Old-fashioned name for helium nucleus (two neutrons, two protons).

ESOTERIC TERMS
(continued)

- *Beta particles*—Old-fashioned name for electrons.

- *Radioactivity*—The emission of alpha, beta, or gamma radiation by an isotope. During alpha or beta emission, one isotope changes into another.

- *Electron*—Negatively charged particle in the atom. Circles nucleus; about 1,800 times lighter than proton and neutron.

- *Proton*—Positively charged particle in nucleus of atom. About 1,800 times heavier than electron but with same size charge.

- *Neutron*—Electrically neutral particle in atomic nucleus. Slightly heavier than proton.

- *Craptron*—Particle found everywhere in Washington, D.C.

- *Nucleus*—Central region of atom where neutrons and protons reside.

- *Atomic mass* (*atomic weight*)—More-or-less number of neutrons plus number of protons in an element. But you must also count energy needed to hold nucleus together.

- *Atomic number*—Number of protons in nucleus.

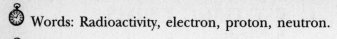

SUMMARY

🕰 Words: Radioactivity, electron, proton, neutron.

🕰 Definition: Radioactivity (see Esoteric Terms).

🕰 Key 'cept: The atom is not indivisible, but composed of smaller particles that were discovered by detaching them from the atom and measuring their charges and masses.

CHAPTER 7

QUANTUM MECHANICS

PHYSICS WALKS
THE PLANCK

YOU MUST REMEMBER THIS

Quantum mechanics—the theory that explains phenomena on the size of atoms—is right. It is also so conceptually weird that physicists to this day feel uncomfortable with it.

BLACKBODIES OVERCOME
CONSERVATIVE PRINCIPLES

Alert readers will have been alarmed at the unprincipled nature of the previous chapter. No new theories, laws, sweeping generalities. Only brave souls groping in the dark with nothing but their noses and their q/m measurements to guide them. Not to worry. This chapter will make up for it. However, it would be a mistake to think that Chapter 6 was completely lawless. All through Thomson's q/m measurements, Newton's laws and Maxwell's equations surreptitiously exerted their force, guiding the next step toward Enlightenment. But in 1900 Newton and Maxwell suddenly met their match with the work of Max Planck—the first quantum mechanic.

The situation was this. In 1634, Father Urbain Grandier of Loudon, France, was condemned as a witch and burned at the stake. He went up in flames. However, not all objects behave this way. As you know, an iron rod thrown into a fire will glow first red, then orange; if the fire is really hot, you might get the thing to glow yellow. The color of the iron rod depends on the temperature. In the late nineteenth century another pastime of physicists was to measure the spectra of various objects. A spectrum, as physicists generally use the term, is a graph of the intensity of light given off at different frequencies or wavelengths. **Intensity, in turn, is defined to be the amount of energy emitted each second per square meter of the body's surface.**[1] If you measured the spectrum for a glowing iron bar at different temperatures you would get results very much like the figure below.

---

[1]Energy emitted per unit time is called *power*, so intensity is power per unit area. One joule per second (energy per unit time) is defined to be one watt. So the units of intensity are watts per square meter.

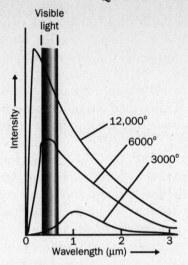

Visible
light

*Blackbody Curves—
These curves show the
intensity of light emitted
at various wavelengths
for three different blackbodies, each at a different
temperature. Tempera-
ture is the only thing that
distinguishes one black-
body from another.*

For historical reasons that aren't too enlightening, the
instantly recognizable shape (which you will never forget) is referred to as a blackbody spectrum. The peak
frequency is the frequency at which most of the radiation
is being given off. For the iron bar in a fire, this would
actually be in the infrared; infrared radiation is sometimes referred to as "heat."

Blackbody spectra are extremely important in physics
because of their universality. **Virtually any hot body emits
electromagnetic radiation in the form of a blackbody
spectrum.** The Sun appears yellow because at 5,800°, the
Sun's surface temperature, the peak of the blackbody
curve lies in the yellow. You also are a pretty good blackbody. If you put your hand under an extremely sensitive
microwave receiver, the receiver would register the blackbody radiation you are emitting.

Hand in hand with the universality of the blackbody
spectrum is the fact that the shape and height of the
curve depend only on the object's temperature. An iron
bar and a steel bar in a fire will produce the same black-

body curve. So will a rock, a lump of clay, a piece of glass, and Urbain Grandier before he catches fire. **Blackbody radiation is totally independent of the blackbody's composition.**

A great puzzle at the turn of the century was to explain the shape of the blackbody curve. All attempts to derive it from the laws of mechanics and electrodynamics failed. The situation was critical.

To get a (very) rough idea why, pretend you are Ravi Shankar playing a sitar inside a cubical room. The sitar is an interesting instrument: You actually touch only certain "main" strings; lying beneath those are a number of "sympathetic" strings that resonate when the main strings are plucked, giving the sitar that otherworldly drone. So, you are playing this sitar. As you pluck the main strings, energy is imparted to the sympathetic strings, which start to vibrate and in turn give off energy as sound waves. The frequency of the sound waves is the same as the frequency at which the strings are vibrating. Long strings give off lower frequencies than short strings (not counting overtones). If we suppose that this is sort of a cosmic sitar with a vast number of sympathetic strings for all lengths, then a vast number of frequencies will be given off.

Now, if as a musician you want to start this cubical room rocking—resonating, to scientists—you try to fill it with "standing waves." You have probably seen such waves when you attach one end of a rope to a wall and shake the other end. You can then create wave patterns that seem to be frozen; the crests and troughs always remain in the same place—"standing." But you can do this only for certain wavelengths. Only waves that can "fit" evenly between the wall and your hand become standing. To "fit" requires a whole number of half-wavelengths. The same is true for sitar strings and for sound waves in a cube.

There is a longest wave that will fit; longer waves are excluded. But there is no shortest wave that will fit; you can always make the wavelengths shorter and shorter in order to set up standing waves in the room. You might then believe that you can fit more short waves in a room than long waves, and this turns out to be true.

Very strange to a *fin de siècle* physicist. The energy you give to the main strings of the sitar gets more or less evenly distributed among all the frequencies given off by the sympathetic strings. This "communism of energy" results from a fundamental theorem in physics called the **equipartition theorem:** energy gets equally distributed among all possible "modes" of vibration. If more higher-frequency modes are allowed in the room, then most of the energy gets concentrated in the high frequencies.

But this is a disaster. In the 1890s, physicists regarded blackbodies as containing a myriad of charged particles that oscillated—like tiny sitar strings. When the charges oscillated, they accelerated and so radiated energy. Suppose, then, you began to heat an iron over a bed of coals, at a few hundred degrees. This infrared energy would get distributed among all the oscillators. But, just as with the sitar in the cubical room, a lot more high frequencies are allowed than low frequencies. Thus, according to equipartition, most of the energy should get converted into higher frequencies and you ought to be blasted from your fireplace by ultraviolet, X, and gamma rays. This is not known to happen.

MAX PLANCK, QUANTUM MECHANIC

The situation was saved in 1900.[2] In an act of desperation, a forty-two-year-old professor, Max Planck, found

---

[2] For more on quantum mechanics, see George Gamow, *Thirty Years That Shook Physics* (New York: Dover, 1985).

that he could fit the shape of the blackbody curve if he assumed that the radiant energy emitted by each oscillator was given by the simple formula

WHO'S

H

O

☞

Max Planck (1858–1947)
German theoretical physicist; introduced the quantum principle into physics. Regretted it forever. Still, one of the greats. One of the few physicists to be alphabetized, by "h," Planck's constant, not mnemonic, but m and p were already taken.

Albert Einstein (1879–1955)
Him again. Introduced the photon, or quantum of light. Also regretted it. Also one of the greats, if you needed to be told.

Niels Bohr (1885–1962)
Applied quantum principle to hydrogen atom, explained hydrogen spectrum. Didn't regret it. Yet another one of the greats.

Louis de Broglie (1892–1987)
Said matter is a wave. Some say he regretted it.

Erwin Schrödinger (1887–1961)
The cat's meow. Invented what is now called quantum mechanics, or wave mechanics. Hated Heisenberg's version.

Werner Heisenberg (1901–1976)
Independently invented a different form of quantum mechanics. Hated Schrödinger's version. Also discovered uncertainty principle, though we can't say for sure.

$$E_{light} = h\nu. \qquad (1)$$

Here ν was the frequency of the light and h was a number. From the experimental data Planck found that h was about 6.6×10^{-34} joule-seconds. Today h is called Planck's constant in his honor. Just as important, he needed to postulate that the blackbody "strings" could not have any old energy, but had to have **discrete energies** given by

$$E_{string} = nh\nu$$

or

$$\nu = E_{string}/nh. \qquad (2)$$

The important thing is that the number n must be an **integer,** 1, 2, 3, . . . Then the bottom form of equation (2) shows that a string (oscillator) with a given energy E_{string} vibrates at a certain frequency when $n = 1$, at another when $n = 2$, etc., but it *cannot* vibrate at frequencies in between. In (2) is contained the original use of the

> **Quantum Principle:** Certain quantities in nature occur only in discrete intervals; the size of these intervals is determined by Planck's constant, h.

Equation (2) also shows that for a given energy E_{string} there is a *highest* possible frequency that a blackbody "string" can emit; it is the frequency given when n has its *smallest* value, 1. That means that the energy you put into an iron in the fireplace cannot get transferred into higher and higher frequencies indefinitely. In this way, Planck solved the "ultraviolet catastrophe," as the transfer of energy to ultra-high frequencies was known. The fact that you do not get fried by gamma rays in front of a fireplace is a direct proof of quantum mechanics. So is the shape of the blackbody spectrum.

PHOTOEFFECT OPENS SUPERMARKET DOORS AND CONFIRMS THE QUANTUM HYPOTHESIS

The idea that energy can only come in discrete intervals is peculiar. It is more than peculiar; it is impossible. Planck tears at his hair, except that he is bald. He spends the rest of his life trying to prove himself wrong. There must be another way to get the blackbody spectrum that conforms to the "classical" physics of Newton and Maxwell. But no, there is no escape. The year is 1905. Bengal is partitioned. The headlines scream:

PHOTOEFFECT EXPLAINED
QUANTUM STRIKES AGAIN
Left-Wing Physicists Argue That Laws of Newton and Maxwell Do Not Govern Nature on Its Most Fundamental Level.

So, the photoeffect. The photoelectric effect, to unabbreviate it, was discovered independently by Heinrich Hertz and Aleksandr Stoletov around 1890 and was another great puzzle of *fin de siècle* physics. Refer to the figure below. Light hits a piece of metal, typically zinc, and the metal ejects electrons, which form an electric current that can be measured with a meter. The photoeffect. Sound simple? Behold:

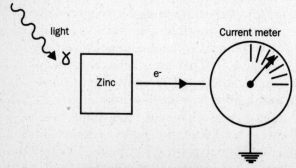

The Photoelectric Effect—
Incoming light hits a piece of zinc to produce an electric current.

1) If the light is ordinary white light, there is no current. In other words, no electrons are ejected, no matter how bright the light.
2) If you increase the frequency of the light to a "threshold frequency," usually in the ultraviolet, electrons are ejected.
3) Above the threshold, the higher the frequency of the light, the more energetic the electrons.
4) Increasing the brightness of the light does not increase the energy of the electrons. It does, however, increase the size of the current.

Confused? You should be. What we call the brightness of a light source is an informal word for its intensity. As already discussed, when you increase the intensity of a lightbulb, you increase its energy output. According to Maxwell, the greater the energy in a light wave, the greater its amplitude. Thus one can think of increasing the intensity as increasing the amplitude of the wave.

In that case, the photoeffect is exceedingly mysterious. No matter how much you increase the energy of white light (by increasing the amplitude of the waves) you can't knock any electrons out of the zinc. But you increase the frequency, which classically *does not* change the energy of the light, and you *do* knock electrons out. Thereupon, you continue to increase the light's frequency and you *do not* increase the number of electrons knocked out, but you *do* increase the energy of the electrons. On the other hand, increasing the brightness, which classically *does* increase the energy of the light, *does not* increase the energy of the electrons, but it does increase the number of ejected electrons. You bet you're confused.

Enter Einstein. Again. In the first of his great 1905 papers, he postulated that Planck's relationship $E = h\nu$ in (1) above does not apply only to light given off by oscillators in a blackbody, but is a fundamental property of light itself. He then took a revolutionary step: He posited that light sometimes acts as particles ("quanta"), each of which carries energy $h\nu$. He then said that the energy of each electron ejected by the zinc is given by the simple formula

$$E_{electron} = h\nu_{light} - E_{threshold}. \qquad (3)$$

Behold: This equation explains everything. $E_{threshold}$ is the amount of energy it takes to knock an electron out of the metal. It is a property of the metal and varies from metal to metal. Now, an incoming light particle strikes the zinc. If its *energy* = $h\nu$ is below the threshold energy, naturally no electrons are emitted and $E_{electron} = 0$. This accounts for observation (1), that light below a certain frequency cannot kick out electrons.

Once $h\nu_{light}$ is above $E_{threshold}$ (which tends to be in the ultraviolet) electrons will be ejected. This explains observation (2). Moreover, increasing ν_{light} increases the electrons' energy $E_{electron}$ in direct proportion (see figure on page 169). This explains observation (3). Finally, instead of associating the intensity of light with the amplitude of a wave, we associate it with the *number* of light particles. Then, if we assume that each light particle kicks out one electron, increasing the intensity increases the number of incoming particles and consequently the number of outgoing electrons. This explains observation (4).

The idea that intensity of light sometimes corresponds to the number of particles, and not to the amplitude of a wave was, as we have indicated, revolutionary but true. By the way, after 1926 Einstein's light quanta became known as **photons,** a term familiar to everyone today.

In 1909 Einstein also postulated (for reasons you don't want to know) that the momentum p of a photon is:

$$p = h\nu/c, \qquad (4)$$

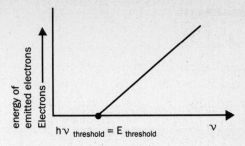

Einstein postulated that the electrons ejected from the zinc have energy $E_e = hg - E_{threshold}$, where hg is the energy of the incoming photons. No electrons are emitted until the photon energy becomes greater than the $E_{threshold}$, which depends on the metal. Thereafter the energy of the electrons is proportional to that of the photons. This explains all the observations above.

where, as usual, ν stands for the frequency. From (1) or (3), this immediately implies

$$E = pc \text{ (for photons)}. \qquad (5)$$

If you don't like this sleight of hand, Equation (5) also follows immediately from (1) and postulate (8) below, using the relationship between frequency and wavelength $c = \lambda\nu$. Not only are (4) and (5) extremely important in their own right, but they allow us to present a reasonably simple derivation of $E = mc^2$ (see below).

WHAT A BOHR

Right-wing physicists do not swallow Einstein's photons. They cling to classical ideas. The year is 1913. Austria creates an independent Albania to block Serbia's outlet to the Adriatic. The *Times* reports:

EAT AT JOE'S
Neon Signs Hold Key to Atom
Prof. Bohr Uses Quantum to Explain Hydrogen Spectrum.

DEMO 1

E = mc² !!!

For this demo you must understand conservation of momentum from Chapter 1. You also need the expression for momentum of a photon from this chapter.

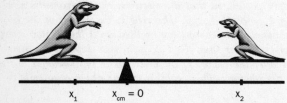

$$x_1 \qquad x_{cm} = 0 \qquad\qquad x_2$$

The heavier creature must sit closer to the center of the seesaw than the lighter one in order to keep it level.

To derive $E = mc^2$ we must first talk about seesaws. Everyone knows that if two creatures of unequal weight are on a seesaw, the heavier one must be nearer the balancing point than the lighter one (figure a).

We can get an exact formula for how far the two creatures must be by considering the *center of mass* ("balancing point") of a system. For two masses, the position of the center of mass (x_{cm}) is given by the equation

$$x_{cm} = (m_1 x_1 + m_2 x_2) \, / \, (m_1 + m_2). \qquad (1)$$

To decipher this, assume that the two masses are equal. Call them *m*. Then (1) becomes

$$x_{cm} = m(x_1 + x_2)/2m = (x_1 + x_2)/2.$$

DEMO 1

(continued)

As you expect, the balancing point is halfway between the two masses. If we put the location of x_{cm} at zero, then $x_1 = -x_2$, which is just another way of saying that the masses are located at equal distances from the balancing point. Indeed, (1) is just the formula for the average distance between the masses.

Now, recall that velocity is the rate of change of position with time. The velocity of the center of mass for a moving system will thus be from (1)

$$v_{cm} = (m_1v_1 + m_2v_2)/(m_1 + m_2). \qquad (2)$$

Notice the terms in the numerator just represent the *momentum* of each mass, $p = mv$. (If you don't remember, see Chapter 1.) Hence the numerator of (2) represents the total momentum of the system. But by conservation of momentum, the momentum of an isolated system is constant. Therefore, **the velocity of the center of mass of an isolated system cannot change.** This is one of the most important consequences of conservation of momentum. In particular, if the total momentum of the system is initially zero, then v_{cm} is zero and *remains zero.* **If the total momentum of an isolated system is zero, the center of mass cannot change position.**

Now we can derive $E = mc^2$. Assume we have two equal masses (M) bolted to a massless seesaw at equal distances from the center of mass (figures b and c).

DEMO 1

(continued)

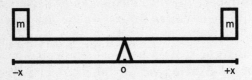

a) The seesaw is balanced with two equal masses (m), one on each end, and the center (or x_{cm}) is at 0. The masses are at equal distances from the center.

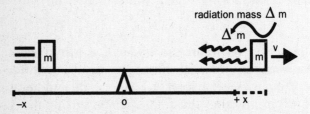

b) If the mass on the right were to give off a burst of radiation toward the left, the entire seesaw would recoil in the opposite direction, much like a skater pushing off a rail.

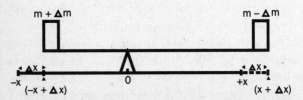

c) When the radiation hits the left mass, the seesaw comes to rest. Its position has shifted to the right an amount Δx, but its center of mass cannot have moved.

DEMO 1

(continued)

We assume the center of mass is located at $x_{cm} = 0$. Then (1) tells us, as before,

$$Mx_1 + mx_2 = 0.$$

Suddenly, the mass to the right gives off a burst of radiation toward the left. By Newton's third law (action and reaction), the whole system must recoil to the right, as in figure c. The radiation will eventually hit the left-hand mass and halt the motion. Before stopping, the system will move a distance Δx to the right. Let's also assume that during the outburst the mass on the right lost some mass, Δm, and the mass on the left gained Δm. But conservation of momentum says that the center of mass of the system cannot have moved. So after the seesaw stops moving, we must have

$$(m + \Delta m)(-x + \Delta x) +$$
$$(m - \Delta m)(x + \Delta x) = 0. \qquad (3)$$

Multiplying all this out and canceling terms gives for the mass loss

$$\Delta m = (\Delta x)\, m/x. \qquad (4)$$

Now let us compute Δx. The distance the masses traveled is just $\Delta x = vt$, where v is the velocity of the seesaw and t its travel time. How long was the system in flight? The length of the seesaw is $2x$, and the velocity of light is c, so the time it took for the radiation to cross the seesaw was nearly $t = 2x/c$ (not counting the distance the seesaw moved while

DEMO 1

(continued)

the light was in transit); thus, very nearly, $\Delta x = 2vx/c$. Plugging Δx into (4) shows

$$\Delta m = 2vm/c. \quad (5)$$

Finally, let's compute v. The radiation carried with it a certain momentum p to the left. By conservation of momentum, the seesaw must have moved to the right with the same momentum p. The mass of the seesaw is $2m$ (not counting the mass carried by the burst of radiation), so its momentum must have been $p = 2mv$. Then (5) becomes

$$\Delta m = p/c. \quad (6)$$

But Einstein showed that the energy E inherent in a pulse of light is just $E = pc$. Consequently $p = E/c$. And (6) then becomes

$$\Delta m = E/c^2.$$

The burst of energy given off by the radiation thus has a mass, Δm, associated with it. If we call the amount of mass exchanged in any process where energy is given off m, we can write this in the familiar form

$$E = mc^2. \quad (7)$$

The famous formula results from conservation of momentum being extended to include energy as well as mass. In deriving (7), we made a couple of approximations, but doing it properly, as Einstein did, leads to the same answer.

Again spectra. Run 10,000 volts through a tube filled with gas and—*voilà*—it lights up: FOOD. If the gas is neon, it appears red. If it's hydrogen, it appears *Miami Vice* pink. Why red? Why pink? Well, take a prism (or better, a 50-cent diffraction grating) and examine the neon sign. You will observe a number of bright red lines. Most of the light is given off at very specific red frequencies (see figure below). Examine the helium sign. It also emits light in bright lines—but not the same colors as the neon. The combination appears pink. Examine a sodium street lamp. Most of the light is emitted in two bright yellow lines. Notice these are *not* continuous blackbody spectra; rather, the light is emitted at particular frequencies—spectral lines. Many gaseous substances when heated emit line spectra. Each substance has a unique pattern of lines. It is exactly this spectral fingerprint that has enabled physicists to decode the composition of the universe. Helium was first identified in 1868 by its spectrum in the solar atmosphere, hence helium's name—from *helios*, the Sun.

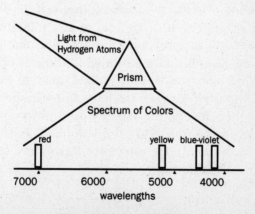

The Hydrogen Spectrum—
A prism breaks up the light from
hydrogen atoms into its separate colors.

As helium indicates, the measurement of line spectra was well underway by the mid-nineteenth century. In 1885 a Swiss schoolteacher, Johann Jakob Balmer, with numerological tendencies, published his one and only scientific paper. He had found, through trial and error, that the lines in the hydrogen spectrum could fit the simple formula, which we write as

$$\nu = cR \left\{ \frac{1}{2^2} - \frac{1}{n^2} \right\}, \quad where \ n = 3, 4, 5, \ldots \quad (6)$$

Here, ν is the frequency of a spectral line, c is the speed of light, and R is just a number (called the Rydberg constant) equal to 1.1×10^7 m^{-1}. To get the frequency of the first hydrogen line, you plug in $n = 3$; to get the frequency of the second, you plug in $n = 4$, and so on.

Balmer had no idea of why this formula worked, but it worked *exactly*. Indeed, new spectral lines were discovered that fitted the rule, so it was difficult to believe that the formula was a coincidence. It was not. Enlightenment awaited 1913. By then Rutherford had already come to regard the hydrogen atom as a miniature solar system, with an electron orbiting a proton like a planet around the Sun. But as mentioned at the end of the last chapter, the electron should radiate away all its energy and fall into the hydrogen nucleus almost instantly. Both Balmer's formula and the stability of the hydrogen atom were explained when Niels Bohr applied the quantum wrench to the atom. He took over Rutherford's picture, adding two postulates:

The Quantization of Angular Momentum: The only allowed orbits are those in which the angular momentum of the electron *mvr* is quantized.

The Quantum Leap: The only time the electron radiates energy is when it jumps between two allowed orbits; it then emits a photon whose energy is equal to the difference in energy between the two.

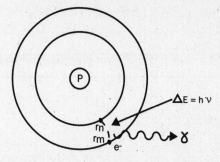

Quantum Leap—
As the electron jumps
from r_n to r_m, it
emits a photon. The
energy level of that
photon is the differ-
ence in the energy be-
tween the two orbits.

Let us elaborate. The first postulate Bohr actually took from an unknown physicist named J. W. Nicholson ("either you do the calculation . . ."). The idea is that, contrary to a planet orbiting the Sun, an electron cannot circle the nucleus on any orbit. Only certain orbits are allowed. Now, you probably didn't notice, but the units of Planck's constant h (joules × seconds) are actually the units of angular momentum (mass × velocity × distance). Thus it was reasonable to postulate that the angular momentum of the allowed orbits should be:

$$mvr = n\hbar. \qquad (7)$$

Here, $\hbar$ ("h-bar") is the standard symbol for Planck's constant divided by 2, or $h/2\pi$. Why 2π? Probably because it worked. As in Planck's hypothesis, n must be an integer. Equation (7) then says that the angular momentum of the electron can only occur in integral multiples of $\hbar$

The second postulate decrees—in total contradiction to Maxwell—that electrons on the allowed orbits do not radiate and by radiating cause the atom to collapse. However, they may magically jump from one orbit to another.

Now, on any orbit an electron has an energy given by the sum of its kinetic energy plus potential energy. The Quantum Leap hypothesis says that when an electron jumps from one orbit to another it emits (or absorbs) a photon whose energy is given by Einstein's formula $E = h\nu$, where now E is equal to the energy difference between the two orbits.

The punch line is that with these two postulates Bohr was able to obtain the Balmer formula exactly! He wrote it as:

$$\nu = \frac{2\pi^2 m_e Z^2 e^4}{h^3} \left\{ \frac{1}{k^2} - \frac{1}{n^2} \right\} \quad (8)$$

Here, m_e and e are the mass and charge of the electron; $1/k$ was the ½ in the Balmer formula and n was the integer. The whole collection of numbers out front was cR, where R was the Rydberg constant. Bohr's reproduction of Balmer's formula was a great triumph for the quantum picture. Moreover, Bohr's formula works for certain atoms other than hydrogen. We see that Equation (8) exhibits the Z^2 dependence that Moseley found in his X-ray spectra, leading him to identify Z as the atomic number. Moseley's work was one of the first confirmations of the Bohr atom.

MATTER MAKES WAVES

Quantum mechanics did not end there. Einstein had postulated that light waves sometimes behave like particles. Could particles sometimes behave like waves? This was the idea of Louis de Broglie, who hypothesized that all particles have a wavelength associated with them given by:

$$\lambda_{db} = h/mv = h/p. \quad (9)$$

In words, the de Broglie wavelength of a particle is equal to Planck's constant divided by the particle's momentum. If a couch potato with mass 100 kilograms is moving at .1 meter per second, Equation (8) shows its de Broglie wavelength is about 10^{-33} meters, so incredibly small as to be unmeasurable. On the other hand, the mass of an electron is only 10^{-30} kilograms and it may be moving at a tenth the speed of light. In this case λ_{db} can be large enough to measure (calculate it!). De Broglie's ideas were confirmed in 1927 when electrons, beamed through crystals, showed wavelike diffraction patterns, just like those in Young's experiment!

The idea that particles sometimes behave like waves and vice versa leads to startling results. For example, we know from Young's experiment that a beam of light shone on two slits will form a specific diffraction pattern on a distant wall or photographic plate (see Chapter 4). This **two-slit diffraction pattern** is conclusive proof that light behaves as a wave and is passing through both slits. If you cover up one slit, you do not get the same pattern.

On the other hand, we have said a beam of light consists of particles called photons. Pretend the beam of light is so dim that one photon at a time is passing by the slits. If photons are particles, surely each passes through only one slit or the other. Indeed, if you perform this experiment you will see on the photographic plate little spots where the photons have fallen, conclusive proof of the particle nature of photons. But as the pattern builds up over many photons, you reproduce the **two-slit diffraction pattern**, which is proof that the light passed through both slits and is a wave! If you cover up one slit or the other each time a photon passes, you won't get the two-slit pattern! Each photon must somehow sense both slits, like a wave, even though it is recorded on the photographic plate like a particle!

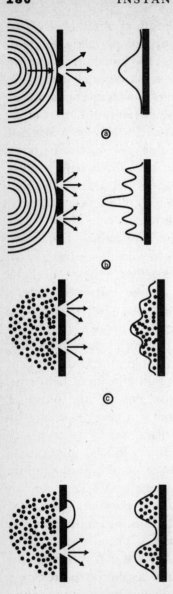

a) *The light wave through this screen creates a one-slit diffraction pattern on the opposite wall.*

b) *The light wave through this screen creates a two-slit diffraction pattern on the opposite wall.*

c) *Sending particle-like photons one at a time through two slits results in a two-slit pattern, after enough photons have fallen on the screen. A particle should only be able to go through one slit, but the two-slit pattern indicates that each photon has somehow gone through both slits, like an extended wave.*

d) *This conclusion is confirmed if one or the other slit is shut at random before each photon passes. Then you get 2 one-slit patterns, not a two-slit pattern. This proves that in (c) each photon particle was nevertheless going through two slits.*

The **wave-particle duality** is the central dilemma of the twentieth century. It has caused untold nervous breakdowns and academic warfare. But just as the True Believer must accept that Vishnu can appear as both turtle and fish, then in order to attain Enlightenment so must one accept that there is no clear distinction between particles and waves. This concept was enshrined by Bohr in his famous

> **Principle of Complementarity:** Waves and particles represent complementary aspects of the same phenomenon.

UNCERTAINTY IS FOR SURE

Most of the developments we have described so far have belonged to what is called "the old quantum mechanics," which combined traditional behavior of particles with the quantum principle. However, the similarity of particles and waves led physicists to invent a new, improved formulation, which became known as "wave mechanics." Principally responsible for the development of wave mechanics were Erwin Schrödinger and Werner Heisenberg. Wave mechanics is a further abstraction from common sense, a quantum leap in abstraction, one might say. Conceptually it is probably the most difficult subject in physics to come to terms with, and debates about what it all means continue among natural philosophers to this day.

In wave mechanics, Newton's equation $F = ma$, which described the motion of an object, is replaced by an equation (the "Schrödinger equation") describing the evolution of a thing called the "wave function." The wave function is generally denoted by ψ (the Greek letter psi), and it is meant to fully describe a particle, but you can't observe ψ directly. Rather, we interpret ψ² as the probability the particle will be found in a certain place at a certain time. What the hell does that mean? For example, we might write down ψ for a photon that passed through the two slits above. It wouldn't tell you where you would find the photon on the wall, but it would tell you the *probability* that a photon would be found in a given place.

You are reeling from the shock of such a staggering idea. The determinism of Newton's second law, $F = ma$, which allowed you to predict the future behavior of an object *exactly* has vanished in a puff of probability! Quantum mechanics is a nondeterministic theory. It does not tell you the outcome of an experiment; it only tells you the odds. It is not the Psychic Friends Network: it is Jimmy the Greek laying odds at Belmont.

Along with the destruction of determinism, quantum mechanics tells you there are other limits to knowledge. We refer to the celebrated

> **Heisenberg Uncertainty Principle:** It is impossible to measure both the position and momentum of a particle with arbitrary precision.

ESOTERIC TERMS
(es-ə-'ter-ik tərms)

- *Classical physics*—The physics of Mozart and Haydn. Any physics that does not have Planck's constant, h, in it. That is, physics before quanta. Not to be confused with the yet-to-be-invented "Physics Classic," in which scien-

ESOTERIC TERMS
(continued)

tific golfers are awarded points for computing the ball's trajectory before it hits the ground.

- *Intensity*—Energy emitted by an object per unit time per unit area. Energy emitted per unit time is defined as power, so intensity is power emitted per unit area. The unit of power is one joule per second, or one watt.

- *Blackbody spectrum*—The characteristic, and nearly universal, spectrum emitted by hot objects. Depends only on temperature, not on composition.

- *Quantum*—The smallest unit of anything. Many politicians seem to have one quantum of intelligence, and zero of fair play.

- *Photoelectric effect*—The effect behind automatic door openers. The ability of photons to eject electrons from certain metals.

- *Photon*—A particle of light.

- *Wave function*—Something that can't be measured but is nevertheless assumed to exist. According to quantum mechanics, it contains all the information about a particle.

- *Quantum mechanics*—The science that nobody can understand. It assumes that certain quantities in nature do not occur continuously, but in discrete intervals.

To be more precise:

$$(\Delta\ p)\ (\Delta\ x) \geq \frac{\hbar}{2}\ .\qquad(10)$$

That is, the uncertainty in any measurement in the momentum of a particle $(\Delta\ p)$ multiplied by the uncertainty in any measurement of position $(\Delta\ x)$ must be greater than or equal $\hbar/2$.

With the exception of the second law of thermodynamics, more nonsense has been written about the Heisenberg uncertainty principle than about any other law of physics. Treat any loose talk of it as you would radioactive waste. Accept (9) as a law of nature. The Heisenberg uncertainty principle and the probabilistic nature of wave mechanics are supremely important because they place on us *limits to knowledge*. It's a sobering thought that we cannot know everything about the universe, but that's the way it is. We will speak a little more about this state of affairs in Chapter 10.

SUMMARY

 Key Words: Quantum, photon.

 Key Definition: $E = h\nu$.

 Key 'cept: Certain quantities in nature occur only in discrete intervals, or quanta. This fact led to a theory—quantum mechanics—that describes nature in terms of probabilities. Quantum mechanics killed the determinism of Newtonian mechanics.

CHAPTER 8

DIVERSITY AND UNITY
STIRRING THE PARTICLE SOUP

YOU MUST REMEMBER THIS

Physics goes through periods when phenomena diversify—and nobody knows what's going on—and periods when phenomena are explained by a unifying principle. This is nowhere clearer than in twentieth-century particle physics.

ART AND SCIENCE

Consider a text on art. Amid glorious reproductions, there will be pointless speculations about the ulterior motives of artists, and unintelligible and even meaningless commentary on artistic trends throughout the ages. You will be exposed to the Italian Renaissance, French Romanticism, Impressionism, Expressionism, Futurism, Cubism, Surrealism, Modernism, and Post-Modernism. By the end, when you gaze at a Brancusi statue or a Gottlieb canvas, you may feel we have not progressed far from Minoan civilization or, worse, you may experience the common reaction that a monkey set loose in a paint shed could have done better.

Contrast your reaction to this book. You may feel the commentary has been unintelligible, but for the most part the laws and discoveries under discussion have had an unalterable meaning. Muddy subjectivism is minimal. Furthermore, there has been clear progress. You found Chapter 7 more difficult to understand than Chapter 1. What's more, what undoubtedly appeared in Chapter 1 to be an amorphous discipline, hardly distinguishable from philosophy, has by Chapter 7 gradually fractured into a number of subfields that seem less based on grandiose philosphical speculations than on experimental results and established laws of nature.

True. The bad news is that science has become so balkanized that specialists can hardly talk to each other, yet alone to the public. The good news is that because diversity and difficulty have crossed some critical threshold, these last chapters will be unconscionable oversimplifications. We will not attempt a derivation. So mix yourself a nonalcoholic diet gin and tonic and relax. Lime, please.

WHO'S

Paul Adrien Maurice Dirac (1902–1984)
Created relativistic quantum mechanics. A positive thinker, he predicted the positron.

Wolfgang Pauli (1900–1958)
Among other accomplishments, he predicted the neutrino.

Julian Schwinger (1918–1994),
Sin-Itiro Tomonaga (1906–1979),
Richard Feynman (1918–1988)
Independently developed quantum electrodynamics, or QED, the theory of how photons and electrons interact. Quite an eminent distinction.

Satyendranath Bose (1894–1974)
Investigated the properties of particles now termed bosons.

Ernest O. Lawrence (1902–1958)
Inventor of the cyclotron. (Had the operator of the machine possessed just one eye, we'd call him "cyclops.")

Hideki Yukawa (1907–1981)
Worked out properties of strong nuclear force; predicted pion.

E.C.G. Sudarshan (1931–   ),
Robert Marshak (1916–1992)
Devised first theory of weak nuclear force. Weak force, strong theory.

Murray Gell-Mann (b. 1929)
Proposed quark theory, thus giving physics an

WHO'S HO☞

(continued)

oddly named family of particles and adding a zany dash of humor to an already nutty profession.

Sheldon Glashow (b. 1932),
Steven Weinberg (b. 1933),
Abdus Salam (b. 1926)
Independently worked out theory combining weak nuclear force and electromagnetic force into one electroweak force, which is how the Energizer bunny feels at the end of a long day.

SUBATOMIC STEW

The unexcelled growth of physics, like some unruly Amazon jungle, is nowhere more apparent than in twentieth-century **particle physics,** sometimes known as **subatomic physics** or **elementary particle physics.** The only thing that saved physicists trying to cut through the tangle of vines and branches was their machete of reductionism: All apparent diversity is the result of a few underlying phenomena.

To set the stage for what is to come, remember that at the turn of the century, physicists knew of only one subatomic particle—the electron. Then came the proton, which was already too many. But that left half the mass of the atom unaccounted for. The answer was a third particle, really an outrage. Everyone of course prayed that the neutron was the end of the story, that all the basic building blocks of nature had been discovered. Nature had other ideas.

One of these ideas was revealed to Paul Adrien Maurice Dirac. You have not forgotten that Maxwell united the fields of electricity and magnetism into one electromagnetic field and in doing so produced the first unified field theory. Since then, physicists have attempted to fit all physics into a single overriding framework. After all, physics is physics and the laws of nature must apply to everything. Quantum mechanics is not allowed to violate relativity and vice versa (see the principle of multilateralism in the Introduction). However, quantum mechanics as originally formulated *did* violate the theory of relativity. The Bohr atom and even Schrödinger's equation used Newtonian concepts that took no account of Einstein's time dilation or mass increase. By the same token, relativity ignored Planck's quantum hypothesis.

Such incompatibility made Dirac unhappy. To cure his spleen, he invented a new unified theory, **relativistic quantum mechanics,** which obeyed all the rules of both special relativity and quantum mechanics. Dirac's theory basically dealt with electrons. But the equation he wrote down seemed also to describe an **anti-electron,** a particle with all the properties of the electron except that it had a positive charge.

In 1933, two years after Dirac predicted the existence of the **anti-electron,** the particle was discovered accidentally by Carl Anderson. It became known as the **positron.** Anderson had discovered the positron in the debris of a **cosmic-ray collision.** "Cosmic ray" sounds like a weapon used by bug-eyed aliens in a bad 1950s sci-fi movie. Actually, it is a vague term for any high-energy charged particle that strikes the atmosphere from outer space. They consist of electrons, positrons, protons, and nuclei of various elements. High-energy gamma rays (photons) also strike the atmosphere but for some reason are not referred to as cosmic rays. When one of these "primary" particles strikes an air molecule, its extraordinary energy can be converted by $E = mc^2$ into many "secondary" and even "tertiary" particles. Anderson's positron was one of these secondary particles. As we shall see, it is in similar manmade collisions that many particles known today were discovered.

The positron is our first example of **antimatter**. Particles like the electron and proton (particles we will later term "fermions") have antimatter counterparts. Matter and antimatter twins are identical except for having opposite charge. As all *Star Trek* fans know, when a particle and an antiparticle meet, they annihilate, converting 100 percent of their mass into energy. Good for a warp drive, but there is a slight problem: where do you find the stuff? Antimatter is generally produced only in cosmic-ray or laboratory collisions. It is a profound fact of nature that the universe does not contain equal amounts of matter and antimatter. That's the way it is.

In 1930, at about the same time Dirac proposed the positron, Wolfgang Pauli proposed the neutrino. This came about as a result of a long-standing crisis: beta decay. As discussed in Chapter 6, since the early 1900s it was known that certain radioactive isotopes decay by emitting beta particles (electrons). But in the 1920s, careful experiments showed that the betas carried away only a fraction of the available energy. Most of it was simply disappearing. The situation is grave. Great outcry and stir:

NIELS BOHR PROPOSES ABANDONMENT
OF CONSERVATION OF ENERGY:
PHYSICS IN DISARRAY

Pauli, in a letter explaining that he could not attend a scientific conference because he planned to attend a gala ball, proposed another out: a new particle that would carry off most of the energy in radioactive decay. This particle was the neutrino. You must realize at the time the neutron itself had not even been discovered, so Pauli's was an extremely radical idea. It turned out to be correct, but this was not directly verified for more than two decades. Why?

Neutrinos are among the most abundant particles in the universe, about as abundant as photons. At any moment untold numbers are passing through your body. But neutrinos are extraordinarily loath to interact with other matter. In fact, they interact so weakly with other matter that the average neutrino would pass through a sheet of lead far thicker than the distance between the sun and the next star! It is no surprise that the neutrino was not actually detected until the mid-1950s. However, the story brings with it a moral: **Thou shalt not abandon the laws of nature.**

THE AVALANCHE BEGINS

With the neutrino and the positron, physicists had discovered the last of the elementary building blocks of nature. Until 1937. In that year a particle 200 times more massive than the electron but otherwise indistinguishable appeared in cosmic-ray debris. The **muon.** As Isador Isaac Rabi once said, "The muon? Who ordered that?" No one has ever satisfactorily answered Rabi's question.

The muon. The last of the fundamental particles. Physicists sigh in relief. Until 1947. The **pion**, about 140 times heavier than the electron, appears in cosmic-ray showers.

Certainly the pion is the last of the elementary particles. Physicists toast themselves for having finally discovered the fundamental building blocks of nature.

The year 1947 moves slowly. Another cosmic-ray shower. Physicists are bewildered. Yet, there is no escape: the **lambda,** heavier than the neutron. The definitely last elementary particle. Physicists pat themselves on the back for having excavated the bedrock level of reality.

It is 1953. Powerful particle accelerators have appeared. Suddenly, the avalanche is unleashed: **kaons, hyperons, sigma plus, sigma minus, cascade particles,** each with an antiparticle. . . . The list soon becomes endless. By the 1960s, physicists have discovered hundreds of elementary particles. How can there be over a hundred elementary building blocks of nature? Physicists cease toasting themselves. They abandon the term "elementary particle physics" and begin to call their field simply "particle physics." In desperation they take over

> **The Totalitarian Principle:** If something is not expressly forbidden, it is compulsory.

Unless there is a law of nature forbidding an event, a phenomenon, a particle, that event, that phenomenon, that particle will exist. With the totalitarian principle comes a bleak corollary:

> **The Experimentalist's Nightmare:** Everything has been predicted.

Indeed, every time an accelerator is turned on, a new particle appears. Someone cracks that a Nobel prize should be given for *not* discovering a new particle.

Taxonomy: Fermions vs. Bosons

But, what . . . what does it all *mean*? That is the question. The answer is written by the principle of reductionism: Simplify! Always simplify! However, before we simplify, we must classify. Attention.

The particles we have spent most of the time talking about, the electron, proton, and neutron, along with the positron and the neutrino, belong to a class of particles termed **fermions,** after Enrico Fermi. Fermions are the particles that make up what we call matter[2]. But the world contains more than matter. It also contains forces and, just as for every thing there is a season, for every force there is a particle.

GEOGRAPHY OF THE SUBATOMIC WORLD

FERMIONS (Matter Particles)		Hadrons	BOSONS (Force Particles)
Baryons			**Mesons** pion, antipion
neutron, n	antineutron, n		π^+ $\bar{\pi}^+$
proton, p	antiproton, p		π^0 $\bar{\pi}^0$
			π^- $\bar{\pi}^-$
Leptons			
muon, μ	antimuon, $\bar{\mu}$		
electron, e^-	positron, e^+		
			Photon
neutrino, ν	antineutrino, $\bar{\nu}$		γ

This chart gives representative particles in each class, together with their Greek symbols. The +, −, or 0 represents their electric charge. The list is by no means exhaustive.

---

[2]Fermions are the particles that have antimatter counterparts: electron-positron, proton-positron, proton-antiproton, neutrino-antineutrino, etc.

To understand vaguely where these force particles come from, let's return to Dirac's relativistic quantum mechanics, his theory of electrons. We know that electrons radiate electromagnetic waves (light) when accelerated in an electromagnetic field. As a physicist you have a burning desire to know the precise details of how the electrons interact with this field and radiate. Unfortunately, Maxwell's theory of electrodynamics is a classical theory (no quantum mechanics). To wed it to Dirac's theory, you need first to quantize the electromagnetic field (introduce Planck's constant somewhere), then combine it with relativistic quantum mechanics to produce a theory of relativistic quantum mechanical electrodynamics.

Such a theory, known by the slightly less ridiculous name of quantum electrodynamics, or QED, was put in final form by Julian Schwinger, Sin-Itiro Tomonaga, and Richard Feynman in 1948. Now, we know that electromagnetic radiation—light—shows up in quantum mechanics as the photon. To sound-bite QED, then, it is the theory that describes the interaction of photons and electrons in a way consistent with both relativity and quantum mechanics.

For us, what is important is the following: QED makes it clear that the photon is also the particle that transmits the electromagnetic force. Indeed, this is the first example of

The Force-Particle Duality: Every force in physics is transmitted by a particle.

Particles that transmit forces are collectively termed **bosons,** after Satyendranath Bose. The photon, then, is a boson. By the above rule, the gravitational force must also have a boson associated with it—a **graviton.** Although such a particle has not yet been discovered, there is evidence for its existence, as we will discuss in Chapter 9.

DEMO 1

Fermions vs. Bosons !!

The world of the physicist is divided into fermions and bosons. But what decides whether a particle is a fermion or a boson? Behold:

The key word is *spin*. In Chapter 7 we found that the angular momentum of an electron in orbit was quantized. That is, its angular momentum could take on only integral multiples of $\hbar$: $mvr = n\hbar$, where n was an integer.

But particles can also have an angular momentum without being in orbit. How is this possible? Well, a top spinning around its axis has an angular momentum. Many particles behave like little spinning tops. Their intrinsic angular momentum is referred to as *spin*. Like orbital angular momentum, a particle's spin can only come in units of $\hbar$. We might write $S = n\hbar$ But here's the rub. The distinguishing trait of fermions is that n is not an integer; it is half an integer, like ½. Indeed the spin of ordinary particles like electrons, protons, and neutrons is $(½)\hbar$. Other fermions have $S = (¾)\hbar$, etc.

Now it is easy to explain bosons. Bosons also have spin $S = n\hbar$, but in their case, n *is* an integer. We have said that a photon is a boson. For a photon, $n = 1$, so its spin is $S = \hbar$. The pion, which transmits the strong nuclear force, has a spin of 0.

FORCING THINGS

There would hardly be any point in introducing a rule if it applied to only one particle, or maybe two. There are, however, more forces than those of electromagnetism and gravity. If you think back to Rutherford's discovery of the atomic nucleus, it is rather strange that a nucleus exists at all. The nucleus is composed of protons and neutrons, the latter being neutral and the former being positively charged. But like charges repel and the coulomb force between two protons 10^{-14} meters apart is enormous. So how is it that the nucleus does not fly apart? There must be an extremely strong force holding it together. Indeed there is, and we term it simply **the strong nuclear force.**

The strong nuclear force was first described by Hideki Yukawa in 1935. He postulated that it was roughly one thousand times stronger than the electromagnetic force, but that it operated only within the atomic nucleus—a very short distance. It is also unlike the electromagnetic force in that it is "charge-blind"; the strong nuclear force works equally well on neutrons as protons. However, consistent with the force-particle rule, the strong nuclear force required its own particle to transmit it. Yukawa postulated such a particle, which was subsequently discovered in cosmic-ray showers. Today it is known as the **pion** and it has a mass 270 times that of the electron.

But the strong nuclear force is not the only nuclear force. About a billion times weaker is the **weak nuclear force.** (How imaginative.) The weak nuclear force is extremely subtle, but without it stars would not shine and life would not exist. So you can't dismiss it. It is responsible for the radioactive decay of nuclei. In particular it shows up in a phenomenon we have already mentioned several times, beta decay, in which a nucleus emits an

DEMO 2

Atom Smashers !

Atom smashers, more properly known as particle accelerators, are the workhorses of the particle physicist. We can get a fairly good idea of how these work from concepts already digested. One can, for instance, simply put a charged particle, like an electron or a proton, in an electric field E. Then from Chapter 4, the force acting on the particle is simply $F = qE$. After accelerating a distance d, the particle will acquire an energy Energy = qEd (energy = force × distance!). The longer the distance, d, the higher the energy.[1]

The accelerator just described is a simple form of the type known as a linear accelerator. Large **linacs** can be several kilometers long. Your television set is a small linac.

Alternatively, one can put a charged particle in a magnetic field and the particle will swing around in a circle. In this case, from Chapter 4, $F = qvB$. Since the particle is moving in a circle, the force on it is the centrifugal force, and as usual, $F_{cent} = mv^2/r$. Equating the two forces gives

$$v = qBr/m. \quad (1)$$

The kinetic energy of the particle, $(\frac{1}{2})mv^2$ is then

$$KE = (\frac{1}{2})mv^2 = q^2B^2r^2/2m. \quad (2)$$

So, the larger you make B or r (the size of the

---

[1]From Demo 4 in Chapter 4, this can also be written as Energy = qV, where V is the voltage. So, the higher the voltage, the higher the energy.

DEMO 2

(continued)

orbit), the more energy you can give to the particles. Accelerators that work by this principle are termed **cyclotrons,** invented by Ernest Lawrence, whom you remember every time you hear the words "Lawrence Livermore Laboratory."

Most modern accelerators are a *bit* more complicated than this description makes them, but you get the idea. Also, to design an accelerator requires that special relativity be taken into account. For instance, from Chapter 5, we know that mass increases with velocity. That means the m in the bottom of (2) gets larger as the particle is accelerated and so KE goes down. To get the same KE you need to make B or r much larger. This is why accelerators cost billions and billions. Einstein is to blame.

electron. The simplest example of beta decay is that of the neutron. A free neutron is actually a radioactive particle, and with a half-life of about ten minutes it decays into a proton *(p)*, an electron *(e)*, and an antineutrino *($\bar{v}$)*:

$$n \rightarrow p + e^- + \bar{v}. \quad (1)$$

The weak nuclear force governs the rate at which this decay occurs. As we just said, the weak force is extremely subtle and was not sorted out until the mid-1950s. The first theory of the weak force was created by E.C.G. Sudarshan and Robert Marshak and, at about the same time, by Feynman and Murray Gell-Mann. Once again, the existence of the weak force demanded a new particle, in this case the W boson (W for "weak").

THOSE QUIRKY QUARKS

What a mess. There are more particles than hyphenated peoples' republics left over from the collapse of the Soviet Union. Can nature really be so malicious? No. But physicists can. Before sorting things out we have to introduce four more terms. Refer to the chart on page 193 to keep from going out of your mind. Greek scholars will have an advantage here.

Baryon comes from the same word as "baritone," meaning "deep" or "heavy." (The *barythron* in Athens was a deep pit into which condemned criminals were thrown.) Baryons are heavy particles that include neutrons and protons. All baryons are fermions but not vice versa.

Meson, as in "mesozoic" or "mezzo soprano," means "middle." Mesons are somewhat lighter particles that include the pion, which transmits the strong nuclear force. All mesons are bosons, but not vice versa.

Leptons are the lightest fermions; they include the electron, muon, and neutrino.

Hadron refers to baryons and mesons together.

Why introduce terminology confusing enough that an economist could love it? Because in 1963 Murray Gell-Mann, a hyphenated person, proposed that the properties of all the known particles could be accounted for if the hadrons were made up of three more fundamental particles he called **quarks**, a name he pulled from James

Joyce's *Finnegan's Wake.* During those years physicists were ravaged by an epidemic of silliness and they named the quarks "up," "down," and "strange." At the time he proposed it, nobody—not even Gell-Mann—believed his theory, but evidence began to accumulate that it was right and physicists became True Believers. It was the ultimate clearance sale: the hundreds of known particles all boiled down to three quarks (and three antiquarks), leptons, and photons.

Along with their strange names, quarks have rather strange properties. For example, if the charge on the electron is taken to be -1, then the charge on the up quark is $+\frac{2}{3}$, and the charge on the down and strange quarks are each $-\frac{1}{3}$. All baryons are made up of three quarks. A proton, for example, is two ups and a down, while a neutron is two downs and an up. All mesons are made of two quarks. The pi$^+$ is an up plus an antidown.

Duck with advanced degree in physics:

Quark! Quark!

SIPRESS

Undoubtedly this all seems complicated. It is actually worse. Since the 1960s, new particles were discovered that required the introduction of three more quarks: bottom, top, and charmed. (For a while the bottom and top quarks were also known as truth and beauty, but bottom and top apparently won out in a return to sobriety.) What's more, all quarks come in "colors": red, green, and blue. These "colors" have nothing to do with ordinary color. It's just another property of quarks and

DEMO 3

Conservation Laws in Subatomic Physics!

The radioactive decay of the neutron is a good illustration of how the laws of physics work on a subatomic level. The neutron is radioactive and decays into a proton, an electron, and an antineutrino with a half-life of about 10 minutes:

$$n \rightarrow p + e^- + \bar{\nu}. \qquad (1)$$

The little bar over a Greek letter denotes an antiparticle; this one is a "nu-bar," an antineutrino.

Note several things. A neutron is electrically neutral. It decays into a proton, which is positive, and an electron, which is negative. The antineutrino is also neutral. So the total charge on the left is zero, as is the total charge on the right. No charge is created or destroyed. **Conservation of charge is obeyed in all subatomic processes.** This is a great help in balancing equations in subatomic physics.

There are two other conservation laws at work in (1). Neutrons and protons are heavy particles, termed **baryons** (see chart on page 193). Electrons and neutrinos are light particles known as **leptons**. Note that the number of baryons on the left of the equation is equal to the number of baryons on the right. This is because, in analogy to conservation of matter, physicists have discovered

Conservation of Baryons: The total number of baryons in any subatomic process does not change.

DEMO 3

(continued)

By the same token, we have

Conservation of Leptons: The total number of leptons in any subatomic process does not change.

Here you might be confused. There are no leptons on the left of (1), but two on the right. Right? No. An electron is a plus lepton but an antineutrino is a minus lepton, so their sum is zero. Antiparticles (including antibaryons) always have to be counted as negative.

another example of silliness on the part of physicists. And since neutrons and protons are made up of quarks, the strong nuclear force between them must in reality be a force between quarks. It is. The quark force is called "glue," and since all forces require a boson to transmit them, there must be a boson that transmits the gluonic force. There is. It is called (you guessed it) the **gluon.** The theory of quarks and gluons is, to make the nonsense complete, known as *quantum chromodynamics*, or QCD.

As we go to press, the last quark—top quark—appears to have been discovered at Fermi lab. One less reason for the Supercollider. The end of physics? Who knows but with six quarks, three colors, and some gluons, QCD really can account for the properties of all known hadrons. A couple dozen particles may still seem like a mess, but it's better than four hundred.

A THEORY OF EVERYTHING?

We have just described how many particles are made of a few particles. This was a great triumph of reductionism. But it was not the end of the line. It turns out that the electromagnetic force and the weak nuclear force had certain things in common. This gave people the idea that maybe they were actually two aspects of the *same* force. In the 1960s Sheldon Glashow, Steven Weinberg, and Abdus Salam developed a scheme in which the electromagnetic and weak nuclear forces became part of one **electroweak** force. The electroweak force required a new particle to go along with the W boson from the weak nuclear force. This new boson was designated the Z^{o}. Both the W and the Z^{o} are very heavy particles—about a hundred times heavier than the proton—and so are very hard to produce in accelerators (Remember $E = mc^2$: The more massive the particle, the more energy you need to produce it.) But in 1973 the existence of the Z^{o} was established and the number of forces in nature was reduced from four to three.

The idea of a unified field theory containing all the forces continues to hold a powerful grip on the minds of physicists. In 1974 Glashow and Howard Georgi proposed that the electromagnetic force, the weak nuclear force, and the strong nuclear force were *all* part of one, overall force. This was the first of the Grand Unified Theories (GUTs), which include all the forces except gravity. As you might guess, GUTs need even more bosons than previous theories. In fact it has twice as many as the electroweak theory. These new bosons have a fatal consequence: protons and neutrons—the very stuff of which we are all made—slowly decay into leptons. Admittedly, this happens over a long time—about 10^{31} years or more—but eventually the entire universe will disintegrate! No more Bach.

It is an extremely depressing thought. On the other hand, huge tanks of water lined with sensitive detectors have been set up around the world to observe proton decay, but so far to no avail. Maybe we are safe after all.

The next step is to incorporate the final force: gravity. Once that task is accomplished physicists will have a "Theory of Everything" (but will it explain the popularity of Madonna?). However, although the media has recently hyped "superstring" theories as the Holy Grail, physicists are less gullible, and it is safe to say that today there does not exist an undisputed theory of **quantum gravodynamics.** So we are saved from talking about it and you are spared the task of memorizing yet another multiconsonant mouthful. But there does exist an extremely successful theory of gravity itself. It is called general relativity and we turn to it now. Pass the Tylenol.

ESOTERIC TERMS
(es-ə-'ter-ik tərms)

- *Antiparticle*—A particle with an opposite charge from its matter counterpart, but otherwise with the same properties.

- *Positron*—An antielectron.

- *Cosmic ray*—Any high-energy charged particle that hits the atmosphere from outer space.

- *Neutrino*—An extremely elusive particle, something like the photon, that is given off in beta decay, for example.

- *Strong nuclear force*—Force that holds the nucleus together. The strongest force in nature.

ESOTERIC TERMS
(continued)

- *Weak nuclear force*—The force that governs the rate of radioactive beta decay, among other things.

- *Pion*—Particle that transmits the strong nuclear force.

- *Fermions*—Particles that make up matter; include protons, neutrons, electrons.

- *Bosons*—Particles that transmit forces; include photon.

- *Baryons*—The heaviest subatomic particles; include neutrons and protons.

- *Mesons*—Middleweight subatomic particle; include the pion.

- *Leptons*—The lightest fermions; include the electron and neutrino.

- *Hadrons*—Baryons and mesons together.

- *Quarks*—The particles that make up the hadrons.

- *Electroweak force*—The force that includes both the weak nuclear force and electromagnetic force.

- *GUTs* (Grand Unified Theories)—As yet unconfirmed theories that combine the strong and electroweak forces, which means theories that include gravity will have to be called Grand Grand Unified Theories. Sheesh.

SUMMARY

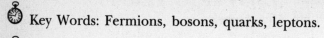 Key Words: Fermions, bosons, quarks, leptons.

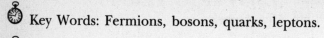 Key Definition: Quarks; see Esoteric Terms.

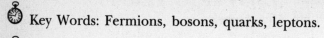 Key 'cept: The principle of reductionism has been successful in explaining the behavior of all known particles in terms of quarks and leptons, and in combining the weak nuclear force with the electromagnetic force.

CHAPTER 9

GENERAL RELATIVITY AND COSMOLOGY

GRAVITY'S BIG DRAW

YOU MUST REMEMBER THIS

Despite all the metaphysical horseshit in the press, the subject of cosmology—the study of the universe at large—is a science, based on the equations of Einstein's general theory of relativity. The Big Bang model of the universe combines general relativity with particle physics and has made enough successful predictions to be believed by everybody but nutcases.

REAL MEN EAT GR

Einstein took about eight weeks to produce special relativity. He took about eight years to produce general relativity. That might give you an idea of their respective difficulties. The bad news is that the mathematics of general relativity is, indeed, *tough*. The good news is that it is so tough that we cannot try to explain it. We can only make analogies and draw a few pictures. Unfortunately, popular explanations of GR (as practitioners call it), which necessarily rely on pictures and imperfect analogies, lead people to believe that GR is more philosophy than physics. General relativity is in truth more old-fashioned than quantum mechanics and particle physics; it had distinct philosophical roots, but as physical theories do, it shed them for mathematical rigor and experimental predictions.

WHO'S WHO ☞

Albert Einstein (1879–1955)
As in the Chapter 5, he's still the main man. Concentrate on Einstein here and you'll be okay.

Like the name **special relativity, general relativity** is something of a misnomer. General relativity is more properly thought of as Einstein's theory of gravity. Its philosophical basis lay in a problem we skimmed over when talking about special relativity in Chapter 5. That burning issue was the question of absolute space. We said that if you had your eyes shut and were on a vibration-free train moving at constant velocity, you could not tell you were moving. This led Einstein to abandon the concept of absolute space, as he stated in his principle of relativity. The fine print here is **constant velocity.** If the train is *not* moving at a constant velocity, but accelerating, you certainly know it; anytime you get a McQuiche attack and floor the accelerator to get you to McQuicheland before passing out, you feel a force pushing you into the car seat. Doesn't that indicate the existence of absolute space? After all, you must be accelerating with respect to *something*. This was Newton's point of view, and for that reason he actually did believe in absolute space, even though when dealing with constant velocities he felt there was no way to detect it.

Ernst Mach (1838–1916) disagreed. He argued that it doesn't make sense to talk about what you can't observe. (Presidents who insist on talking about economic recoveries, take note.) The only thing we have a right to do is say that we feel forces when we eat quiche or accelerate with respect to the distant stars. This idea became known as

> **Mach's Principle:** An object at rest or moving at constant velocity with respect to the distant stars feels no forces acting on it.
>
> Conversely, an object accelerating with respect to the distant stars feels forces acting on it.

This is Mach's reformulation of Newton's first law, the law of inertia.

WHAT'S GRAVITY GOT TO DO WITH IT?

If you reenter Chapter 5, you will see that special relativity had a great number of successes. But all these triumphs rested on that caveat **constant velocity.** The word **acceleration** does not even appear in Chapter 5. The fact that special relativity did not take into account accelerations disturbed Einstein greatly. With his principle of relativity he had thrown the idea of absolute space onto the ash heap of history. Would the existence of acceleration force him to resurrect it? Would he be forced to abandon his carefully forged principles? No! Mach's principle gave him faith; absolute space does not exist, accelerations be damned. But how to incorporate Mach's principle into a theory of relativity that would include accelerations? Einstein is depressed. He scratches his head. He plays the violin. He becomes increasingly desperate. He eats quiche.

Then, in 1907, he had what he called "the happiest thought of my life." The idea was a simple one. Imagine you are depressed over the failure of your recent diet and decide to commit suicide by jumping down an elevator shaft at the office. You have taken your pet frog with you. During your long fall, you release it. But it continues to croak in your face because, as Galileo discovered three hundred-odd years ago, all objects fall at the same rate, regardless of their mass. The frog in your open hand feels weightless to you because it is not pressing down on your palm.

You have also taken your hated bathroom scale along with you in order to destroy it. But to your delight, it registers your weight as zero pounds. You have quantified the observation of your frog: falling objects are weightless. Of course, you have little time to savor your success. . . .

But let us restate the moral so it is crystal clear: objects falling freely under the acceleration of gravity feel no gravitational force. This is why astronauts are weightless in orbit, not because gravity has turned off above the atmosphere or because they are far from the Earth; they are falling freely around the Earth at the same rate as the space shuttle.

To continue on your suicidal plunge: On your way down you are met by an elevator going up. The meeting is abrupt; you suffer fifty-three broken bones and internal hemorrhaging. Nevertheless, as you accelerate upward, you notice that the bathroom scale now indicates your weight is far higher than usual. Damn. . . .

Still, you have increased the fund of scientific knowledge: The accelerating elevator is making you heavier; it is effectively producing a gravitational force. You begin to hallucinate: you imagine you are trapped in a windowless elevator. You feel weightless. Does that mean you are far from the Earth's gravity, or are you merely freely falling in the Earth's gravitational field? Now you are suddenly pressed to the floor. Are you accelerating? Or have you merely come to rest in a box on the ground and experiencing the Earth's gravitational pull? In your last seconds of lucidity you arrive at a blinding conclusion: There is no way you can tell the difference between an acceleration and a gravitational force. Then all is dark.

This insight is what Einstein called the "happiest thought of his life." He named it the

> **Principle of Equivalence:** Accelerations are indistinguishable from gravitational fields. They are equivalent.

In the case of free-fall, your downward acceleration is just enough to cancel the gravitational force. Then the principle of equivalence immediately implies what you found en route to suicide: Free-fall is equivalent to the absence of gravity. Remember: "Falling-free or orbiting 'round, equivalence says gravity not found."

Einstein was modest calling equivalence a principle. You might get the idea that it is easily abandoned, like "the principle of democracy." In fact, it is a law of nature and it has been by now tested to extraordinary precision. No experiment has ever been found to violate it.

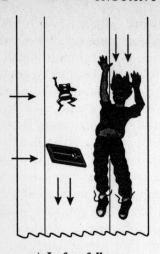 

a) In free fall, you, your pet frog, and your bathroom scale will all fall at the same rate. If you step on the scale it will register zero and you will rightly conclude you are weightless.

b) According to the principle of equivalence, the acceleration of the rising elevator creates a gravitational force of its own. You now weigh more than you had in the first place.

But why was Einstein so excited about what appears to be a pretty obvious idea? Once again, special relativity lacked any mention of accelerations. Through the principle of equivalence, Einstein realized that to incorporate accelerations into relativity he was going to have to create a theory of gravity. This is the theory the world knows as general relativity, Einstein's theory of gravitation.

I CAN SEE MY DEFLECTION!

Once Einstein had the principle of equivalence (PoE), he needed only eight more years to work out the details. But the PoE alone was enough to allow him to make the most famous prediction of GR: that light is deflected in a gravitational field. This is easiest to understand with the help of a few diagrams, so if you spend a few minutes with the figure on page 214, you will see that if you shine a light beam across an upwardly accelerating elevator, you will naturally enough see the beam bend downward. By the equivalence of acceleration and gravity, the same thing must happen in a gravitational field. *Voilà*.

Einstein completed GR in 1915. Four years later, after an interruption by World War I, Arthur Eddington led a famous expedition to Africa to determine whether light was deflected by the gravitational field of the Sun. The idea was to measure star positions during a total eclipse and compare these positions to those on photographic plates taken when the Sun was absent. GR predicts that starlight grazing the Sun should be deflected by 1.75 seconds of arc, about the angle made by the body of Madonna at a distance of nearly two hundred kilometers. A small effect indeed.

But large enough to make Einstein famous. When Eddington reported that his eclipse results verified Einstein's theory the news rocketed around the world.

NEWTONIAN IDEAS OVERTHROWN.
WE DON'T KNOW WHAT IT'S ABOUT,
BUT WARPED SPACE SOUNDS GOOD.

Actually, to measure light deflection is a difficult job, and in the wake of Eddington's expedition, many physicists have questioned whether his experiment was actually good enough to get the right answer. As one noted relativist has said, "If the answer could have been anywhere from zero to infinity, and you ask yourself what were the odds Eddington could have gotten 1.75 seconds of arc without knowing the result beforehand, you realize that there was a lot of wishful thinking going on." So it goes.

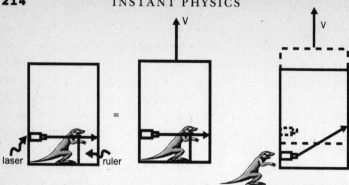

a) A creature shoots a bullet across an elevator. The bullet remains at a constant height above the floor.

b) According to both Newton and special relativity, there is no difference between an elevator at rest or one moving upward at a constant velocity, v. Thus the creature doesn't know whether it is in an elevator at rest or one moving at velocity v.

c) If this bothers you, imagine the scene as viewed by a creature outside the elevator. The bullet has not only a horizontal velocity but has the same vertical velocity as the elevator. So the bullet appears to travel on a diagonal. However, since it has moved upward the same amount as the elevator, it strikes the opposite wall at the same height above the floor as it began.

In any case, Einstein didn't care. As one of his acquaintances told your author in an oft-repeated tale, "When the eclipse results were announced, I ran down the street to Einstein's house in a great agitation and said, 'Einstein, Einstein, Eddington has confirmed your theory!' Einstein shrugged and replied very calmly, 'It would have been too bad for God had I been wrong.'" Regardless of Eddington's results, subsequent experiments have all shown that God was on Einstein's side.

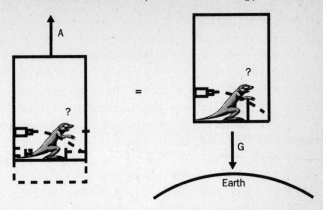

d) *Not true for acceleration. The bullet is shot out with a vertical velocity, v, which it maintains. But the elevator is accelerating. So by the time the bullet hits the far wall, the elevator has risen **farther** than the bullet. A creature on the inside will see the bullet falling downward.*

e) *According to the principle of equivalence, the same must hold in a gravitational field and the same is true for light as for bullets.*

WARPED IN SPACE

Where did this tiny angle, 1.75 seconds of arc, come from? Magic? Well, sort of. Knowing what you do already, you can estimate that the deflection must be *something* like that (see Demo 1). Indeed, Newton could have gotten half the correct answer. But to get *exactly* the right answer does take a bit of magic: you must accept what GR tells you—that space is curved. Indeed, because space and time in both theories of relativity are combined into one, you must accept that not only space, but spacetime is curved. Like a banana.

This is an exotic idea but by now everybody believes it. Suppose two rebel scout ships take off on parallel courses in deep space. What happens? As you know, according to the Euclidean geometry you were forced to swallow in high school, parallel lines do not intersect, so the scout ships just continue traveling on their parallel lines forever. Suppose, instead of scout ships, we now have two imperial battle cruisers—real massive monsters—and we start them off on parallel lines. What happens? Well, after a while their mutual gravity pulls them together until they collide. Darth Vader bites the dust.

The difference in the two cases is shown in figure 1. In the first case, we can say that the scout ships' gravity was negligible, since they didn't collide and their trajectories remained parallel. In the second case, we might say the battle cruisers' gravity was not negligible because initially parallel lines finally intersected. However, there is an alternate way to phrase what happened. Parallel lines remain parallel only in Euclidean geometry, the geometry that takes place on a flat piece of paper—flat space. But on a globe—curved space—initially parallel lines (like lines of longitude) *do* intersect. You can say that gravity pulled the battle cruisers together, or you might say that they were traveling in curved space. Thus, according to the PoE and GR

$$\text{free-fall} = \text{absence of gravity} = \text{flat space}$$

$$\text{acceleration} = \text{gravity} = \text{curved space}.$$

The equations of GR describe exactly how the presence of matter warps space. It takes an awful lot of matter to curve space appreciably. The Sun curves space only enough so that light is bent from a straight path by that minuscule angle 1.75 seconds of arc. Around more massive stars, the angle is bigger. Around galaxies it is bigger yet. The deflection of light around galaxies is large

DEMO 1

Light Deflection Around the Sun !!

You need to know the law of gravity from Chapter 1, the definitions of force and acceleration.

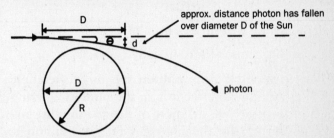

approx. distance photon has fallen over diameter D of the Sun

It is not difficult, using the basic Newtonian physics you have already mastered, to estimate the amount light will be deflected when passing by the Sun. We try to find the angle θ in the figure. Assume a photon streaks by the limb of the Sun, as in the figure. Let's pretend it behaves like a bullet traveling at the speed of light, c. (This is okay; if the photon has energy, then it has an equivalent mass by $E = mc^2$.) The Sun exerts a gravitational force on it given by our friend the law of gravity:

$$F_{grav} = ma = GMm/R^2 \qquad (1)$$

where M and R are the mass and radius of the Sun, and m is the "mass" of the photon. So the m's cancel out and the downward acceleration of the photon is just

$$a = GM/R^2. \qquad (2)$$

If the photon accelerates downward for a time t at this acceleration, it will travel a vertical distance

DEMO 1

(continued)

given by the famous formula (Equation 8, Chapter 1),

$$d = (\tfrac{1}{2})at^2 \qquad (3)$$

or, plugging *a* from (2) into (3),

$$d = (\tfrac{1}{2})GMt^2/R^2. \qquad (4)$$

What is the time *t* the photon spends accelerating? The gravitational attraction on the photon is strongest nearest the Sun, so most of the action must take place during the time it takes for the photon to cross the Sun, or $t = D/c$, where *D* is the Sun's diameter (time is distance divided by velocity!). Substituting this expression for *t* into (4) gives

$$d = (\tfrac{1}{2})GMD^2/(c^2R^2). \qquad (4)$$

Now, the deflection angle θ (in the scientific unit of radians) is very nearly *d/D*. (If you are not familiar with radians, just accept this.) Thus we get θ merely by dividing (4) by *D*, or

$$\theta = d/D = (\tfrac{1}{2})GMD/(c^2R^2). \qquad (5)$$

But the diameter is just twice the radius, $D = 2R$, so (5) becomes

$$\theta = GM/c^2R. \qquad (6)$$

To get the real answer of general relativity you need to multiply (6) by 4. But we haven't done too badly for an estimate. When you plug in the numbers for *M, G, c,* and *R,* you get about 8.5×10^{-6} radians, or 1.75 seconds of arc, the angle that made Einstein famous. Magic.

enough that astronomers have observed a dozen or so examples of **gravitational lensing.** This is a situation in which light from a distant astronomical object is bent around an intervening galaxy so that we see two or more images of it!

The existence of gravitational lenses is only one of many confirmations of GR—and curved space. Of course, the most dramatic example of curved space is found around black holes. These fabled objects are formed when massive stars finish their nuclear burning, lose the pressure holding them up and collapse under their own weight. In the process, their gravitational field increases so much that any light entering them cannot escape, hence, "black hole." Due to the fact that they are invisible, they are rather difficult to observe; black holes can only be detected through their gravitational

a) Two scout ships start off on parallel paths and never meet. You might conclude they were traveling in flat space, where parallel lines never intersect.

b) Two battle cruisers begin on parallel paths but eventually crash. You might conclude gravity was pulling them together, or you might conclude they were traveling on a curved space, where parallel lines meet.

c) Curved space, not fate, pulled these two ships together. On a globe, parallel lines actually meet.

DEMO 2

Doppler Shifts and the Expansion of the Universe !

In 1845 Buys Ballot placed a brass band on a train and observed that, as Christian Doppler had predicted, the pitch of the trumpets rose as the train approached and decreased as it receded. Special relativity predicts that as a result of the time dilation, the same "Doppler shift" takes place for light when a light source is moving relative to an observer. That is, if a stationary lamp emits light of a certain frequency ("pitch"), that frequency goes *up* when the lamp approaches and *down* when the lamp recedes. In the former case we say the light is *blueshifted* (since blue light is on the high-frequency end of the visible spectrum) and in the latter case we say the light is *redshifted.*

The Doppler effect has proven to be of the utmost importance in astronomy. The precise spectrum emitted by an astronomical object tells us what elements it contains. For this reason, spectra have been the key to decoding the composition of the universe. But the Doppler shift gives us more: We can also tell how fast a given galaxy is moving toward or away from us. Observations in the early part of the century showed that most galaxies are redshifted, and hence moving away from us. Indeed, in 1929 Edwin Hubble found a simple relation ("Hubble's law") between galactic redshifts and distances that could only be interpreted as meaning that all galaxies are moving *away from each other.* It is for this reason that astronomers believe the universe is expanding.

effects on nearby objects. We content ourselves with pointing out that the Hubble Space Telescope recently confirmed the existance of a giant black hole in the center of another galaxy, and astronomers have several excellent candidates for smaller black holes.

We might mention at this point one other prediction of GR not connected with the bending of light. Just as Maxwell's theory of electromagnetism predicts that accelerated charges radiate electromagnetic waves, Einstein's general relativity predicts that accelerated masses should emit gravitational waves. Gravitational waves are like electromagnetic waves in that they travel at the speed of light. They are also like electromagnetic waves in that there should be a **graviton**—a particle analogous to the photon—that is the particle version of these waves and that transmits the gravitational force. But gravitational waves are unlike electromagnetic waves in that they can not be detected by a radio. Rather, if a gravitational wave passes by your house, it would squeeze it slightly in one direction and stretch it in another.

Not to worry. Gravitational waves are too feeble to trigger the next California earthquake. To get an idea of how feeble, wave your arm. Don't be shy. Observe, it is accelerating. According to what we just said, it is emitting energy in the form of gravitational radiation. Since $E = mc^2$ this energy is carrying off mass. But this diet would not be approved by the Food and Drug Administration. To test its effectiveness, wave your arm for about 10^{64} years (far longer than doomsday). You might lose one gram of fat through the emission of gravitational radiation.

Indeed, gravitational radiation is so weak that you'd need a star collapsing into a black hole or a supernova to produce enough to be detected in an earthbound laboratory. At the present time, to be honest, they haven't been detected. Physicists are currently building a facility called LIGO—for Laser Interferometric Gravitational Observatory—with which they hope to observe gravity waves. Nevertheless, there is already good experimental reason to believe they exist. Two stars circling each other are accelerating and, like your arm, emitting gravitational radiation. As the stars lose energy, they gradually spiral into each other. Astronomers have at their disposal a particular double-star known as the **binary pulsar,** which allows them to monitor exactly how fast the stars are spiraling into one another. The observed rate is so closely in accord with GR predictions (to about 14 decimal places) that one may argue relativity has now been more accurately tested than quantum electrodynamics!

THE BIG PICTURE

Gravity is a romantic force for another reason. The strong nuclear force—the strongest force of all—does not operate outside the nucleus of atoms. Neither does the weak nuclear force. The electromagnetic force, about 10^{39} times stronger than gravity, does not operate between electrically neutral objects. Planets, stars, galaxies are electrically neutral. Therefore it is left to gravity, the weakest force of all, to determine the fate of the entire universe. General relativity, then, is the theory that describes the evolution of the universe at large.

ASK MR. LIZARD

The study of the evolution and large-scale features of the universe is known as **cosmology,** often confused by government agencies with **cosmetology,** the study of makeup and nail polish. In the last thirty years cosmologists have devoted most of their energy to sorting out this confusion, and in their spare time studying the very beginning of the universe, a silent event popularly known as the Big Bang. Because this topic blends in with astronomy, it does not lie on the central artery of our book. Nevertheless, since the Big Bang theory incorporates so much of what we have discussed and has received so much popular attention, it would be foolish to ignore it completely.

Let us concede immediately that we do not know where the Big Bang came from. There may have been a Big Crunch before it, or maybe the Big Bang just popped out of nothing. Such questions are very murky and cosmology may never be able to answer them. Obviously, when talking about the beginning of *everything* one gets into uncomfortable philosophical–religious quandaries.

We will ignore them. Nevertheless, despite uncertainties about the Moment of Creation, cosmologists have combined general relativity and particle physics to create a reasonably complete picture of the early universe. Here is a fast-food outline of the subject with which you may attack infidels or *New York Times* reporters who claim the Big Bang never happened. The main evidence for the Big Bang is:

1) The expansion of the universe. Astronomers observe that galaxies are moving apart from each other, making it a fair presumption that at one time in the past the matter in the universe was much more crowded together. By projecting the motion of galaxies backward, one can estimate that the moment of maximum compression—the Big Bang—occurred about 15 billion years ago.

2) The cosmic microwave background radiation (CMBR). Astronomers observe that the universe is giving off blackbody radiation like a "hot" object at 3° above absolute zero. As the universe expands, the CMBR gets cooler. Coupled with observation (1), this implies that the universe was much hotter in the past.

3) The abundances of the light elements. Astronomers observe that about 25 percent of the universe by mass is concentrated in the element helium, and lesser amounts in some of the other light isotopes. The observed abundances are exactly what one would expect if these isotopes were "cooked" by nuclear fusion in a hot universe during the first ten minutes after the Big Bang.

4) The bumps in the cosmic microwave background. In the early 1990s, the Cosmic Background Explorer (COBE) satellite observed bumps in the CMBR, which represent the beginnings of galaxy formation.

That's all there is to it. What, not satisfied with fast food? Not filling enough for you? Leaves you asking, "Where's the beef?" Very well, more nourishment in each of the four food groups can be found in the accompanying features.

We end this excursion into GR by reemphasizing a point made in Chapter 8. The theory of general relativity "looks" very different mathematically from the other theories we have discussed. This has made it extremely difficult to construct a "Theory of Everything," which unifies gravity with the three other forces of nature. Claims of superstrings aside, we can say that the feat has so far eluded physicists. On the other hand, such a Grand Unified Theory would probably operate only in the first 10^{-43} seconds after the Big Bang. This is much less than a wink of God's

DEMO 3

Cosmic Background Radiation !

The expansion of the universe allows for the possibility that at some time in the past that expansion began. This moment is the Big Bang. But is there any direct evidence? There is. At least pretty direct. It is called the cosmic microwave background radiation (CMBR), and was predicted in 1948 by Ralph Alpher and Robert Herman, and discovered by Arno Penzias and Robert Wilson in 1964.

Penzias and Wilson at Bell Labs were using an antenna built for satellite communication to do some radio astronomy. Much to their surprise they detected a faint signal whose intensity was unchanging regardless of the time of day or the direction they pointed the antenna. It seemed to be coming from *everywhere*. Recall from Chapter 6 that all hot objects give off blackbody radiation. For a while, Penzias and Wilson thought the mysterious signal was emitted by bird shit in the antenna.[1] A group of Princeton cosmologists had another explanation: This was blackbody radiation given off by the entire universe.

Twenty-five years of observations have proved the Princeton group right: the universe is giving off black-body radiation as if it is a "hot" object at 3° above absolute zero. This puts the peak of the curve in the microwave part of the spectrum. In 1989–1990 the Cosmic Background Explorer satellite (COBE) observed the CMBR to unprecedented

---

[1] See Steven Weinberg, *The First Three Minutes* (New York: Basic Books, 1977).

DEMO 3

(continued)

accuracy and it turned out to be the most nearly blackbody spectrum ever measured.

The only reasonable way to interpret the CMBR is that it is heat left over from the Big Bang. As the universe expands, it cools off. Its temperature is now at only 3° above absolute zero. But in the past, it was extremely hot. Hot enough to fry . . . elements.

COBE also found tiny variations in the temperature of this blackbody curve. Such temperature bumps represent places where matter was just starting to clump together at the beginning of the universe. The discovery was an important confirmation of how galaxies would form in the Big Bang model.

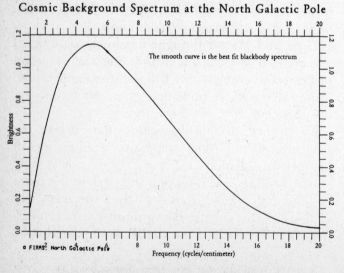

Cosmic Background Spectrum at the North Galactic Pole

The smooth curve is the best fit blackbody spectrum

Brightness

a FIRAS: North Galactic Pole

Frequency (cycles/centimeter)

COBE (Blackbody Spectrum Taken by Satellite).

DEMO 3

(continued)

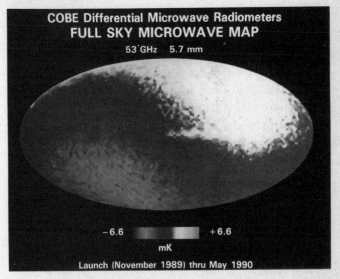

COBE Differential Microwave Radiometers
FULL SKY MICROWAVE MAP
53 GHz   5.7 mm

−6.6 ▬▬▬ +6.6
mK

Launch (November 1989) thru May 1990

COBE Map with Temperature Fluctuations—
Tiny temperature fluctuations in the cosmic background radia-
tion dating from 300,000 years after the Big Bang were discov-
ered by COBE. They represent the beginnings of superclusters
of galaxies.

eye. So you might wonder what all the fuss is about. After
all, according to the fundamental principle of science, theories must be tested by experiment. How could such a theory be tested? The answer is: your guess is as good as mine.
And with that, we leave the universe to its own devices,
since things have clearly gone out of sight.

DEMO 4

The Light Isotope Abundances !

If one goes back about 15 billion years, to the first few minutes after the Big Bang, the temperature was high enough so that nuclear fusion of neutrons and protons could take place (Chapter 6). The process is similar to what takes place in the center of the Sun. In "less time than it takes to boil a potato" protons (p) and neutrons (n) fused into helium via the reaction chain shown in the figure below.

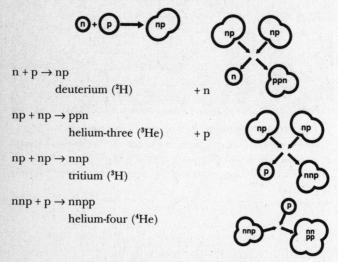

$n + p \rightarrow np$
    deuterium (^{2}H)   + n

$np + np \rightarrow ppn$
    helium-three (^{3}He)   + p

$np + np \rightarrow nnp$
    tritium (^{3}H)

$nnp + p \rightarrow nnpp$
    helium-four (^{4}He)

The Primordial Nucleosynthesis Chain

The figure on page 229 shows the results of computer simulations. At about 3 minutes after the Big Bang, the amount of helium begins to increase rapidly. After about 10 minutes, most of the reactions have stopped

DEMO 4

(continued)

and roughly 25 percent of the universe by mass has been baked into ordinary helium (^{4}He). The graph shows the abundances of the other light isotopes. Astronomical observations confirm *all* these abundances, a great piece of evidence that the universe went through an element-cooking stage right after the Big Bang.

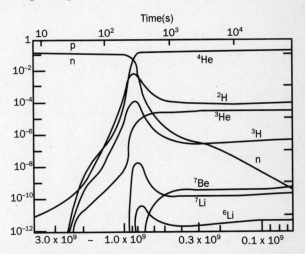

Results of the primordial nucleosynthesis chain. The graph shows what fraction of the total mass ended up in various elements as the temperature of the universe dropped. As the time scale shows (top), the amounts stabilized within about 1,000 seconds after the Big Bang.

SUMMARY

 Key Word: Gravity.

 Key 'cept: Principle of equivalence; acceleration is indistinguishable from gravitational field; to put acceleration into special relativity requires a theory of gravity; gravity in turn curves space.

CHAPTER 10

THE END

YOU MUST REMEMBER THIS

If you ask a stupid question, you may feel stupid; if you don't ask a stupid question, you remain stupid.

OVERSIGHTS

With general relativity and cosmology, we come to the end of our survey of physics. You have attained Enlightenment.

Actually, your author is blushing at what fell from the Path. There has been no mention, for instance, of astrophysics, the application of physics to the workings of stars, galaxies, quasars, and other exotic astronomical objects. But astrophysics is indeed an application, not a fundamental science. The physics that governs stars and quasars is the same physics that governs a fluorescent lamp, only the circumstances differ.

Much the same can be said about solid-state physics. This is the branch of physics responsible for such practical devices as the transistor, superconducting magnets, electronic Ginsu knives, and computers. Indeed, probably a majority of physicists make their living as solid-state physicists. And you thought natural philosophers were impervious to financial gain. . . . No, the main reason solid-state physics was ignored, apart from space, was that it is a somewhat advanced topic. But again, it is based firmly on the principles we have discussed (especially those in quantum mechanics), and if you corner the nearest solid-state physicist and bombard it with stupid questions, you should be able to attain the sacred solid-state knowledge.

Another field that fell from the Path is one that has gotten considerable media attention over the past decade: it goes by the unfortunate name of chaos. The word "chaos" implies randomness and chance—like in throwing dice—which isn't quite what's going on here. Recall we ended Chapter 1 by pointing out that Newton's laws were deterministic. If you knew the initial positions and velocities of all the particles in the universe, then with $F = ma$ you could predict the future behavior of the entire cosmos for all time to come. A tacit assumption of physicists for the past three centuries was that if you then varied the initial conditions just a tiny bit, the behavior of the system would differ only a *tiny* amount from what it had been before. It's a lie.

The equation $F = ma$ was, is, and always will be deterministic. There is no chance involved, no built-in random behavior. But in the past few decades, physicists have discovered innumerable systems whose future behavior is nevertheless unpredictable. These systems are not necessarily very complicated either. It might be a model solar system consisting of only the Sun and Jupiter. A comet swings in from afar. Will it be captured by the Sun? Will it swoop out again into deep space? Will it orbit Jupiter? Depending on the comet's initial direction and velocity, it might do any of these things—or behave in a totally erratic manner that is impossible to predict by classical physics.

All the objects in this solar system—the Sun, Jupiter, the comet—obey $F = ma$. The system is deterministic. But its behavior is infinitely sensitive to the initial conditions. Vary the initial trajectory of the comet in the thousandth decimal place—and all hell breaks loose. The comet's path becomes totally different from what it was before and impossible to forecast. Such systems are said to exhibit **chaos;** a better word might be **deterministic chaos.** And as already mentioned, many such systems are now known. The real solar system is now known to be chaotic in the long run. The toy Chinese acrobat on your author's desk, who attempts to loop over a horizontal bar, the weather . . . the list is endless.[1]

The discovery of chaos has shed new wisdom on phys-

---

[1]There are many books on chaos. James Gleick's *Chaos* concentrates more on the personalities than the science. For those more interested in science than personalities, try David Ruelle's *Chance and Chaos* (Princeton: Princeton University Press, 1991).

ics and philosophy. The universe is not a giant clockwork after all. Left to its own devices, it will start ticking erratically, maybe pop a spring. One can't tell exactly because the rules governing chaotic behavior, like the Heisenberg uncertainty principle, put an absolute limit on knowledge. We can never know the future, so you may stop dialing your 900-number astrologer now. It is also possible that the second law of thermodynamics—the increase in entropy—is related to deterministic chaos. To rigorously establish that connection would be a tremendous advance toward Nirvana.

THE END IS NIGH

So it seems that new discoveries keep coming and there is yet work to be done. On the other hand, from time to time notable physicists—most recently Stephen Hawking and Steven Weinberg—claim that the end of physics is in sight. Is it?

One can't deny that physics has changed. We almost certainly know most of the laws of nature (and now you do too) and they are unlikely to be modified unless it is in conditions approaching those of the Big Bang itself. In terms of fundamental laws, what's left for adepts may indeed be a mop up operation. However, when discussing the end of physics, one should make clear what one means by "end." It seems physicists have several distinguishable ends in mind when this question arises at cocktail parties.

There is first the **In Principle End** of physics. We have a Theory of Everything, which contains all the known interactions. We have attained Nirvana, the going-out of the flame. Physics is over. Yet, the Nirvana End assumes the **Computational End** of physics. That is, it assumes that your Theory of Everything is simple enough to allow you to calculate all the things you want to calculate and design experiments to test it. But so far, Theory-of-Everything candidates are so complicated that nothing whatsoever can be calculated with them. So even if the Nirvana of physics is nigh, it will probably not have much practical impact.

The Tomb of the Unknown Quantity

In contrast to the In Principle, or Nirvana, End of physics, there is the **Stupidity End** of physics. The human brain may be built in such a way that it cannot comprehend the basic structure of the universe. The Stupidity End seems quite likely.

The Stupidity End of physics may, however, be pre-empted by the **Electronic End** to physics. As soon as we construct computers that are smarter than we are, they will take over research and leave the rest of us to watch MTV.

We should also distinguish the **Metaphysical End** of physics. The fundamental principle of science is that theories of quantum gravity and the like are so far beyond conceivable experimental tests that they may forever be relegated to the kingdom of speculation. If we have reached the Metaphysical End of physics, physicists will have become the modern equivalents of medieval theologians who counted angels on pins and the perfections of God.

Immediately threatening, of course, is the **Funding End** of physics. Governments may well decide that physics is no longer worth pursuing and that will make all the other ends irrelevant.

So, in order to discuss whether the end of physics has arrived, one must pick the proper end. We leave that and the subsequent debate up to you. Oh, but let's not forget the **Publisher's End** to physics. All books must be limited to 237 pages and therefore

---

COCKTAIL PARTY CONVERSATION

A physicist is wearing a "Physics Is All There Is" T-shirt. You, wearing a "Cosmology Takes GUTS" T-shirt, approach.

YOU: Do you really believe your T-shirt?

PHYSICIST: Of course. Everything is based on physics. Everything stems from the interactions of the fundamental particles through the four known forces.

YOU: And once you have a Theory of Everything you will explain everything?

PHYSICIST: That's the idea.

YOU: Like the four forces?

PHYSICIST: Absolutely.

YOU: Like why you like bagels?

PHYSICIST: Well—

YOU: Or why people fall in love?

PHYSICIST: Well, in principle it's all got to be a result of physics.

YOU: Isn't it going to be a little *tough* for a Theory of Everything to calculate why people fall in love?

COCKTAIL PARTY CONVERSATION

(*continued*)

PHYSICIST: That's just a practical difficulty.

YOU: Is it? Don't things like chaos and the Heisenberg uncertainty principle put a theoretical limit on what you can calculate?

PHYSICIST: That is true—

YOU: So how can you even consider calculating why you like a bagel? Isn't this in principle impossible?

PHYSICIST: Well, um, I'm not sure. Probably.

YOU: So your Theory of Everything can't explain why I prefer cream cheese, no lox—

PHYSICIST: But—

YOU: —or why Central Europe has fragmented, or why Andy Warhol got even 15 minutes, or why the NRA always misquotes the Second Amendment, or the sound of mountain streams, or—

PHYSICIST: Okay, okay, maybe it's a Theory of Almost Everything—

PHYSICIST: A Theory of a Few Things—

YOU: —or why junk sells, or—

PHYSICIST (moving off): A Theory of a Couple Things?

YOU: —or why terrorist activities—

PHYSICIST (continuing to move away): A Theory of More Than Nothing? A Semi-Grand Unified Theory? An Impressive Attempt to Unify Physics? A Failed Attempt at Advertising Physics? An Excuse to Stop Doing Physics? The Theory That Will End Funding. . . .

YOU: A bagel and cream cheese, please.